Correcting the Blueprint of Life

Correcting the Blueprint of Life

AN HISTORICAL ACCOUNT
OF THE DISCOVERY
OF DNA REPAIR MECHANISMS

Errol C. Friedberg
*University of Texas Southwestern
Medical Center at Dallas*

Cold Spring Harbor Laboratory Press 1997

Correcting the Blueprint of Life

AN HISTORICAL ACCOUNT
OF THE DISCOVERY
OF DNA REPAIR MECHANISMS

© 1997 Cold Spring Harbor Laboratory Press
All rights reserved
Design by Emily Harste

Library of Congress Cataloging in Publication Data

Friedberg, Errol C.
    Correcting the blueprint of life : an historical account of the discovery of DNA repair mechanisms / Errol C. Friedberg.
        p. cm.
    Includes bibliographical references and index.
    ISBN 0-87969-507-2
    1. DNA repair--Research--History. I. Title.
QH467.F748 1997
574.87'3282--dc20   96-45981
                                                                                                                                         CIP

Authorization to photocopy items for internal or personal use, or the internal or personal use of specific clients, is granted by Cold Spring Harbor Laboratory Press for libraries and other users registered with the Copyright Clearance Center (CCC) Transactional Reporting Service, provided that the base fee of $0.25 per page is paid directly to CCC, 222 Rosewood Dr., Danvers, MA 01923. [0-87969-507-2/97 $0 + .25]. This consent does not extend to other kinds of copying, such as copying for general distribution, for advertising or promotional purposes, for creating new collective works, or for resale.

All Cold Spring Harbor Laboratory Press publications may be ordered directly from Cold Spring Harbor Laboratory Press, 10 Skyline Drive, Plainview, New York 11803-2500. Phone: 1-800-843-4388 in Continental U.S. and Canada. All other locations: (516) 349-1930. FAX: (516) 349-1946. E-mail: cshpress@cshl.org. For a complete catalog of all Cold Spring Harbor Laboratory Press publications, visit our World Wide Web Site http://www.cshl.org/

*To my schoolteachers, who gave me a love of history
and the international DNA repair community,
whose collegiality and support I have
enjoyed for so many years*

Contents

PREFACE *ix*

CHAPTER 1   *In the Beginning*   1

CHAPTER 2   *Let There Be Light: The Discovery of Enzymatic Photoreactivation*   27

CHAPTER 3   *The Emergence of Excision Repair*   63

CHAPTER 4   *New Mechanisms for Repairing DNA*   113

CHAPTER 5   *Mismatched Bases Are Repaired by Yet a Different Mode*   139

CHAPTER 6   *The Remarkable SOS Phenomenon*   161

CHAPTER 7   *Epilogue: Molecule of the Year 1994*   185

NOTES *191*

INDEX *203*

Preface

THIS BOOK WAS PROMPTED by a number of intellectual interests, influences, and concerns that more or less coalesced for me several years ago. One of these reflects my growing interest in history in general. I have often pondered why some of us are more taken with events of the past than are others. Speaking for myself, experiencing almost any historically venerated place has evoked a palpable sense of excitement and awe as long as I can remember. So, it is probably not purely coincidental that history was one of the few subjects to which I fully applied myself at school. But the notion of dedicating myself to a formal training in history never entered my head, and this book was most certainly written from the perspectives of a professional historian. Regardless, I believe that as a practicing scientist I have benefited from my general knowledge about the subject matter at hand, and while I am confident that a professional historian could have identified the essential elements of the discovery of DNA repair mechanisms, I am equally certain that it would have taken him or her considerably more time and effort to sort out areas of relative priority and to identify the principal players. Perhaps the ideal scientific history should reflect the collaborative efforts of a knowledgeable scientist and a competent historian.

My interest in the history of the discovery of DNA repair emerged from my long working association with this discipline of biology, in particular writing several texts on the topic. While engaged in those tasks, my curiosity was often piqued about the precise historical events surrounding a number of important discoveries. However, such digressions were not appropriate for a descriptive textbook, so I made the conscious decision to research some of the history of DNA repair and mutagenesis as a separate work. As I began to document a list of scientists whose contributions I knew would be indispensable to this effort, another catalyst to action surfaced. Truth is of course inevitably distorted by forgotten nuances, inaccurate recollections, and interpretative biases, even when recounted by the best intentioned primary sources. But history becomes even more confounded when its primary makers are no longer available for comment, and I was alarmed to discover how many notable contributors to the early history of DNA repair were deceased at the time that I initiated this project. I therefore recognized the imperative of attempting to capture the thoughts and recollections of as many living early contributors to the field as I could reasonably identify.

I was also motivated to write this book by the rapid advances that the field of DNA repair has witnessed in recent years. The manifold responses of living cells to genomic insult and injury now impinge tangibly on numerous other aspects of cellular metabolism, especially oncogenesis, cell cycle regulation, transcription, and DNA replication. I believe that this is an opportune time to document the period when DNA repair and mutagenesis existed more or less outside the mainstream of molecular biology, so that future historians can examine this modern period of synthesis.

Finally, and perhaps most importantly from my perspective as a teacher and educator, it is my view that the startling progress the biological sciences have enjoyed over the past 40 years or so has been so rapid that most, if not all, students of biology have had little, if any, time to appreciate how rich and significant the events of the second half of the 20th century have been. It is my sense that many of them are familiar with the cliché that all practicing scientists stand on the shoulders of legions of our predecessors. However, my experience is that few students truly recognize how many and how broad these shoulders are. Many students are also unaware of how much the actual practice of science has changed in the past 50 years. Biological research has presumably always been extraordinarily interesting, compelling, and consuming to those fully drawn to understanding its deeply locked secrets. No doubt the heady sweetness of unlocking even the tiniest of these secrets has been a creative driving force that has transcended every period of discovery. But the discipline of biology was not always as crowded with scientists as it is now, and the imperatives for success were not always as intensely demanding and as competitive. Hence, I hope that in these pages students beginning their scientific careers will discover not only the particular pathways by which the DNA repair field blossomed and flowered, but also how intensely charming and pleasurable traveling those pathways once was for many of their predecessors.

Historically, the topic of DNA repair emerged from, and is still inextricably interwoven with, aspects of the disciplines of photobiology and radiobiology. This book does not attempt to trace these important historical relationships. That would require much more dedicated study and would likely emerge as a much larger work. My primary intention was to document the history of the *discovery* of the more important presently known basic mechanisms by which cells respond to DNA damage, with a primary emphasis on the discoveries themselves, rather than the events that followed and subsequently brought these discoveries to full maturation. The book was intentionally constructed to be relatively brief. To achieve this goal, I have assumed the reader has some knowledge of basic molecular biology and some familiarity with aspects of DNA repair. My hope is that the established community of DNA "repairologists" will find this book informative, revealing, and sometimes amusing. I hope that other students of the genome and of biology in general will also find nuggets of enjoyment in these pages. In particular, I believe that the book will provide a solid and readable introduction for students about to embark on an advanced undergraduate or graduate level course on DNA repair and mutagenesis.

This work is based primarily on interviews conducted in person or by telephone. In this regard, I offer my sincere thanks to John Cairns, John Clark, Jim Cleaver, Raymond Devoret, Renato Dulbecco, Maury Fox, Sol Goodgal, Philip Hanawalt, Bob Haynes, Adelyn Kelner, Sandy Lacks, Tomas Lindahl, Matt Meselson, Paul Modrich, Miroslav Radman, Stan Rupert, Leona Samson, Dick Setlow, Evelyn Waldstein, and Evelyn Witkin for their generous contributions of time, effort, and patience. All these individuals also provided invaluable written information by way of correspondence, as did Dick Boyce, Francis Crick, Larry Grossman, Walter Harm, Bob Painter, Emmie Riklis, Jane Setlow, Bernie Strauss, Mutsuo Sekiguchi, Frank Stahl, Paul Swenson, Dick von Borstel, and Jim Watson. I trust that I have recorded their recollections, views, and opinions accurately. Particular thanks are due to John Cairns, Phil Hanawalt, Bob Haynes, Dick Setlow, and Evelyn Witkin, and especially my colleagues, Lisiane Meira and Michael Reagan, who labored meticulously through innumerable drafts of many of the chapters, and to Phil Hanawalt, Bob Haynes, and Dick Setlow for providing the initial motivation and encouragement to undertake this task, a task that each of them might have done better justice to. I also wish to thank John Inglis of Cold Spring Harbor Laboratory Press for his interest in publishing this work; Catriona Simpson for her outstanding editing; and Jan Argentine, Denise Weiss, and Maryliz Dickerson for their invaluable help in bringing the book to completion. Finally, generous thanks are due to Glenda Westerfield for her conscientious job in retrieving innumerable articles and books from various libraries.

The French writer Voltaire once wrote that "all our ancient history . . . is no more than accepted fiction." I trust that the history of the discovery of DNA repair mechanisms is not too ancient to have degenerated to this state. However, I have no doubt that other scientific contributors to this field and more than a few historians may have different views of how events unfolded in the past half century and of their relative historical significance. If so, I hope that they will be motivated to document their own version of these events so that future historians will be able to sift through richer mines of source material. Finally, I hope that in focusing on some personalities to the inevitable exclusion of others, I have not made too many enemies among my peers and colleagues, and that I may even have made a few friends.

ERROL C. FRIEDBERG

March 1997

I HAVE CONTENDED that scientists have a responsibility because of their understanding of science, and of those problems of society in which science is involved closely, to help their fellow citizens to understand, by explaining to them what their own understanding of these problems is.

Linus Pauling to Horace Judson
The Eighth Day of Creation

WHAT CONSTITUTES an acceptable explanation in science? As so often, it's hard to see from the light into the dark: the apparent brilliance of present understanding hides from us what went before.

Horace Judson
The Eighth Day of Creation

CHAPTER 1
In the Beginning

IN HIS HISTORICAL ACCOUNTING of the emergence and efflorescence of molecular biology in the mid 20th century entitled *The Eighth Day of Creation: Makers of the Revolution in Biology*, Horace Judson reminds us that "mutations obtained one way or another have . . . always been the chief tool of experimental genetics." But beginning with Mendel, biologists engaged in exploring the nature of the gene and of inheritance were heavily reliant on the appearance of *spontaneous* alterations in gene function in their chosen experimental subjects, be they sweet peas, humans, or, as popularized by Thomas Hunt Morgan, the fruit fly *Drosophila*. Spontaneous mutations are of course infrequent events and the ability to recognize them in the mirror of altered phenotypes was further hindered by the prevalent predilection for organisms with diploid genomes as experimental subjects. Hence, the discovery in 1927 by Hermann J. Muller that X rays greatly enhanced mutation frequency in *Drosophila* (a finding that earned him the Nobel Prize 19 years later) represented a major contribution to genetics. The enormous experimental utility of the increased mutation frequency attendant on the exposure of organisms to X rays was clearly underscored by Muller himself in a paper that he presented at the 1941 Cold Spring Harbor Symposium, a scientific event that we shall shortly revisit.

> The inordinate effectiveness of X-rays and related radiation . . . has hardly been realized to the full. For when we say that a dose of 110,000 r-units results in about 100 times as many lethals per X-chromosome in *Drosophila* spermatozoa as ordinarily occur in untreated material, we do not take into consideration the fact that, in most experiments, all these mutations were produced by a treatment lasting an hour or less, whereas the "natural" mutations represent the accumulation of a whole *Drosophila* generation, that is, of two weeks or more. When this time difference is taken into the reckoning, we find that, during the time of treatment here in question, not 100 but at least 35,000 times as many mutations were produced in the treated as in the untreated material; moreover, by the use of the tubes of higher radiation output now available, this rate of production could be stepped up almost indefinitely.

The ability to manipulate the function of genes with X rays was rapidly broadened to include ultraviolet (UV) radiation and eventually chemicals. UV radiation ultimately proved not only to be a more potent mutagen than X rays for many studies, but also a highly selective probe of the gene itself in living cells. Had investigators paid greater attention to the selectivity of short wavelength UV radiation for nucleic acids relative to other chemical components in the cell, our understanding of the chemical nature of the gene might have been significantly accelerated. Ultraviolet radiation was discovered in 1801 by J. W. Ritter, who found that the blackening of silver chloride by light was greater in the dark region beyond the violet end of the light spectrum than anywhere else in the spectrum. Recognition of the utility of UV radiation as a physical probe to explore the biology of living cells dates back more than a century. As early as 1877, Downes and Blunt reported to the Royal Society of London that sunlight killed bacteria, and they showed that this effect was chiefly associated with the short wavelength component of the radiation. But as the noted photobiologist John Jagger has pointed out, although UV radiation was quickly recognized as safer and easier to use than ionizing radiation:

> Its very availability is deceiving, for it encouraged some researchers to do extensive experimentation with little knowledge of the tool at hand. As a result, there are papers in the literature, dealing with interesting problems, in which the techniques are so inadequate that many of the results are unacceptable. *The proper conduct of ultraviolet experiments in biology is at least as difficult as for X-ray experiments, and the proper interpretation at least as subtle.*

The impact on genetic studies hastened by the discovery that X rays and UV radiation can interact with and in some way alter the "hereditary material" of cells is beyond the province of this book. Of singular interest to us in this treatise is the fact that these discoveries spawned the development of a new and distinctive investigative focus in the field, namely the perturbation of genes by exogenous physical agents, and the scrutiny of the ways in which living cells respond to such perturbations. This focus emerged fleetingly as early as the mid nineteen-thirties, but its maturation to the intellectually comprehensive body of knowledge that we now refer to as DNA repair and mutagenesis was sporadic and slow. The field of DNA repair did not attain full experimental clarity until the late nineteen-forties, and only achieved formal recognition as a distinctive biological phenomenon in the late nineteen-fifties. Why was this era in the history of genetics, an era that was characterized by the extensive, almost routine, use of exogenous agents known to alter genes, not immediately followed by investigations about the nature of these alterations and their physiological consequences? To pose the question another way, what cultural and intellectual influences delayed the emergence and maturation of the notion that genes can sustain repairable damage?

Before attempting to answer these questions, it is instructional to steep ourselves more deeply in the general intellectual climate that characterized the science of genetics in the late nineteen-thirties and early nineteen-forties. A fitting place to begin is the 1941 Cold Spring Harbor symposium on *Genes and Chromosomes, Structure and Organization*, where Muller so authoritatively anointed the impact of the mutagenic properties of X rays on experimental genetics.

The Biological Laboratory at Cold Spring Harbor on Long Island, New York has long been one of the major intellectual breeding grounds of modern genetics and molecular biology. The famous Cold Spring Harbor Symposia on Quantitative Biology were formally initiated in 1933 by the director of the Laboratory at that time, Reginald G. Harris, and with a brief interruption during World War II, these have continued on an annual basis, covering a wide array of topics of biological interest. In his introductory remarks which opened the first symposium entitled *Surface Phenomena*, Harris implored the 29 assembled conferees to:

> Give special consideration to theoretical and controversial aspects, that the discussion may be both significant and creative, and that these conferences may be of the greatest possible value not only to those of us who take part in them, but also to those who will have occasion to refer to them.

Prior to 1941, the symposia lasted for as long as five weeks and the majority of the participants remained in residence in the Laboratory for at least part of that time. Thus began the tradition of convening at Cold Spring Harbor for the summer, a tradition that attracted geneticists and microbiologists from around the country and even abroad. These visiting scientists often brought their own critical pieces of equipment and favored experimental subjects.

The published proceedings of the symposia are important historical archives of the progress of various subdisciplines in biology. Aside from the written accounts of papers delivered at the meetings, the symposium volumes are distinguished by a comprehensive recounting of the discussions that followed each formal scientific presentation and, in more recent years, by a series of candid photographs of "key" conferees. I have never determined who decides which photographs are included in the published volumes. But having attended several such meetings, I have on occasion had cause for wry amusement in watching the subtle jockeying of some of the younger conferees when the symposium photographer is in attendance, presumably in the hope of being captured for posterity standing next to, or better yet actually talking to, some scientific luminary. Nonetheless, paging through some of the older volumes and putting faces to names now legendary in the annals of biology is an enchanting exercise. It is equally interesting to peruse the list of symposium participants included in each of the published volumes, regardless of whether or not they presented formal talks at the meeting.

The 1941 symposium was graced by the presence, among others, of Max Delbrück and Salvador Luria, whose direct and indirect influence on the field of DNA repair will be recounted in several places in this book. Hermann Muller, Barbara McClintock, Milislav Demerec, (the director of the Laboratory at that time, about whom we shall hear more in the next chapter), Alexander Hollaender, Hans Ris, Alfred Mirsky, Curt Stern, Lewis Stadler, and Daniel Mazia were among other scientific notables in attendance. The general stage on which the drama of genetics was unfolding in the nineteen-forties was eloquently summarized by Neville Symonds in some of his historical reflections written in 1988. Symonds reminds us that by 1940, studies carried out principally in maize and *Drosophila* had established the essential foundations of genetics. It was known that inheritance was determined by genes that were present on chromosomes. It was known too that chromosomes were duplicated in all cells by a process of mitosis, except those in the germ line, where they underwent meiosis. Mutations were known to reflect occasional spontaneous changes in genes, and their frequency could be impressively increased by exposure of cells to radiation. Enzymes were known to exist in cells and the nucleus was known to be rich in nucleic acids, but there was a very strong "feeling" that genes were made of proteins. As Symonds put it, "the foundations of formal genetics were well laid, but there were no real ideas about biochemical genetics."

According to Richard Kimball, a longtime contributor to the DNA repair field and one of the conferees at the 1941 meeting, the symposium was characterized by considerable controversy about the mutagenic properties of ionizing radiation. The prevailing model of genes as "beads on a string" lent itself to the notion that mutations and chromosomal aberrations could arise from breaks caused by ionizing radiation. However, the experimental limitations of distinguishing between mutations resulting from deletions and rearrangements of genes, and those possibly arising from chemical alterations of the gene itself, were profound. In his 1987 review entitled "The Development of Ideas about the Effect of DNA Repair on the Induction of Gene Mutations and Chromosomal Aberrations by Radiation and Chemicals," Kimball suggests:

> The difficulty arose in distinguishing between true gene mutations, taken to mean an alteration in the structure of the gene itself, and such breakage-and-reunion events as losses of single genes and changes in gene action by movement to new positions (position effects) . . . . Thus there was no real basis for any discussion of the role of repair of radiation damage to the gene itself.

Nonetheless, the observation that chromosomal breakage induced by X rays was sometimes accompanied by aberrant chromosomal rejoining, which resulted in translocations, did not go unnoticed. Though not documented in the formal paradigm of gene repair, these events were discussed and considered by some in the general context of restitution phenomena. Evelyn Witkin, whose venerable contributions to the DNA repair and mutagenesis field are discussed in later chapters, was a young investigator at Cold Spring Harbor in the early nineteen-forties. She avowed that, in retrospect, her comprehension of the observations on chromosome rejoining by Berwin Kauffman and Barbara McClintock at Cold Spring Harbor in the early nineteen-forties sensitized her to the notion of the rectification of genetic damage, though she did not translate this notion to that of "gene repair" until many years later.

But in general, in the early nineteen-forties, insightful dialogue about mechanisms of mutation induction and the nature of cellular responses to genetic insult was suffocated primarily by the informational vacuum about the chemical nature of the gene itself. For this was a time not yet illuminated by either the 1944 paper of Oswald Avery, Colin MacLeod, and Maclyn McCarty showing that the genetic material was in fact DNA, nor by Watson and Crick's elucidation of the structure of DNA. The prevailing dogma was that genes were made of proteins. This dogma was so entrenched in mainstream biology that according to Judson, when James Watson arrived in England in 1951 to begin the fellowship that culminated in his revolutionary observations with Francis Crick a full seven years after Avery, MacLeod, and McCarty's famous paper, most biologists still firmly believed it. Interred in the complex history of this dogma one can identify a powerful abstraction that was antithetical to the notion of the repair of genes. Quite simply, proteins were believed to be highly stable biological macromolecules. They were not considered as mediators of genetic alteration and were certainly not thought of as substrates for biochemical reactions that mediated the repair of genetic material.

Many of the issues surrounding this dogma are recounted in *The Eighth Day of Creation*. Judson informs us that Oswald Avery was not a geneticist bent on solving the chemical basis of heredity. He was a physician trained as an immunologist and bacteriologist who worked at the Rockefeller Institute and was intent on understanding the phenomenon of pneumococcal transformation established in the late nineteen-twenties by the pioneering studies of Frederick Griffith. Griffith, also a physician, had documented the phenotypic conversion (transformation) of a particular serological type (type II) of the bacterium *Streptococcus pneumonia* (a nonvirulent form of the organism in mice), to a different serological type (type III) which was virulent in mice. He achieved this transformation by inoculating mice with viable type II bacteria mixed with type III organisms that had been inactivated by heat and hence were incapable of growth. Not only were the mice killed by this inoculum, but the recovered organisms acquired the serological properties of the type III form of *pneumococcus*. Most intriguingly, this transformation to the type III form was permanent, apparently representing a fixed heredible change. Avery and his colleagues extended Griffith's observations by showing that this transformation required neither mice nor heat-inactivated type III cells. The same result could be achieved by simply growing type II cultures in a test tube in the presence of a filtered cell-free extract of type III organisms. So, in effect, Avery and his team devised an in vitro system for pneumococcal transformation. They went on meticulously and laboriously to purify the "transforming principle" to apparent physical homogeneity and showed that it was composed exclusively of DNA.

Much has been made of the fact that the classic paper of Avery, MacLeod, and McCarty was published in the *Journal of Experimental Medicine*, the "in-house" journal of the Rockefeller Institute, rather than in a fundamental genetics journal where it might have had more immediate exposure to geneticists. Nonetheless, this paper secured its proper place in history because it represents the first documented demonstration that genes (the transforming principle) are made of DNA. Although they were willing to risk the assumption that "the sodium desoxyribonucleic acid and the active principle are one and the same substance," Avery and his coauthors qualified this statement with the caveat that:

> It is of course, possible that the biological activity of the substance described is not an inherent property of the nucleic acid but is due to minute amounts of some other substance adsorbed to it . . .

Judson recounts that not the least of the many solicitous influences at work at that time was the fact that the Rockefeller Institute was still smarting from an earlier scientific faux pas. Richard Willstatter, also at the Rockefeller and one of the preeminent organic chemists of his age, had years earlier claimed that he could show enzymatic activity in protein-free solutions. This claim was soundly rebuffed by John Dexter Northrop's demonstration that Willstatter's experimental preparations were contaminated with very small amounts of active enzyme.

This historical anecdote concerning the work of Colin Avery and his collaborators illuminates the profound entrenchment of the "protein dogma" concerning the chemistry of genes. A less celebrated but equally illuminating example of this bias can be found in a paper presented by Alexander Hollaender and his colleague C.W. Emmons at the 1941 Cold Spring Harbor Symposium. Hollaender, one of the leading figures in the then infant discipline of radiobiology, may justly be acknowledged as the founding father of the DNA repair field. Born in Germany in 1898, Hollaender emigrated to the United States in 1921. He attended Washington University in St. Louis and obtained his Ph.D. in physical chemistry from the University of Wisconsin in 1931. In a paper written in 1986, entitled "History of Radiation Biology from a Personal Point of View," an expansion of an address presented three years earlier when he received the Enrico Fermi Award, Hollaender wrote that:

> The early nineteen-thirties were an exciting time for those who were interested in modern physics, and I hardly need to mention many of the important scientific developments by Niels Bohr, or by Drs. Heisenberg, Pauli, Schrödinger, Dirac, Wigner, and many others, especially the Berkeley group who considerably influenced those of us who had come into biology from physics or physical chemistry. I had many interesting discussions with Warren Weaver on this most important development. He emphasized time and time again that the biological effects of radiation would become very important, especially since physical and chemical approaches can be used extensively. It was Weaver who first coined the term "molecular biology."

In the early nineteen-thirties, Hollaender was prompted by Warren Weaver to survey the literature and write a report for the Rockefeller Foundation on the biological effects of radiation. He systematically evaluated close to 5000 papers in the field, a task that ultimately provided significant impetus for the establishment of financial support for radiation research by the Foundation. "No other source of support for research of this type was available during the Depression," Hollaender wrote. Financial support for radiation biology was to increase in a massive way in the following decade. During the Manhattan project that led to the development of the atomic bomb, laboratories dedicated to this effort were built in various parts of the country, including Argonne near Chicago, Los Alamos in New Mexico, and Oak Ridge in Tennessee. Following World War II and Franklin Roosevelt's death, President Harry Truman established these and other facilities as permanent National Laboratories for radiation research under the aegis of the Atomic Energy Commission. The focused mission of these National Laboratories contributed to their evolution as centers of excellence in radiobiology and they became the breeding grounds for many prominent contributors to the field of DNA repair and mutagenesis. The nuclear priorities of the U.S. government being what they were at that time, these laboratories and the programs cultivated within their walls were lavishly funded by research standards. This situation did little to endear the community of radiobiologists to the "aristocrats" of molecular biology, who not only labored under more restrictive financial conditions, but additionally considered much of the biological research done in these Laboratories as frankly pedestrian. Regardless of the validity of these judgments about radiation research and the true motives for their expression, there is little question that this mentality was a prevalent leitmotif in the early days of molecular biology (and to some extent still is). It almost certainly contributed to a general climate of disdain for the study of the physiological responses to DNA damage (or at least for the manner in which many classical radiobiologists tackled this problem).

The notion held by many biologists that radiobiological research was in general inferior in its rigor and quality to mainstream molecular biology was not really justified. Each of the National Laboratories evolved divisions expressly dedicated to investigating the biological effects of radiation, many of which supported some excellent research programs. Alex Hollaender spent much of his professional career as director of the Biology Division of the Oak Ridge National Laboratory, an organization that under his leadership came to enjoy a considerable reputation. The Oak Ridge laboratories eventually included a scientific staff of several hundred, including Richard B. Setlow, Jane Setlow, William Kimball, Elliot Volkin, Fred Bollum, Sheldon Wolff, and R. C. (Jack) von Borstel, all of whom made important contributions to genetics and molecular biology. Hollaender was elected to the National Academy of Sciences in 1957 and, as mentioned above, received the prestigious Enrico Fermi Award from the U.S. Department of Energy in 1983. Dick Setlow recalls that he first met Alex in 1955 when he [Hollaender] visited Yale University, where Setlow was working. "We chatted at the time and I remember him telling me that I needed to learn more biology and that I never could learn it at Yale, but if I came to Oak Ridge they would teach it to me," recalled Setlow. "I laughed inwardly at his presumption, but he was right." As we shall see in a later chapter, Setlow did indeed move to Oak Ridge, a change in his career that turned out to be most providential.

Very soon after he assumed the directorship of the Biology Division at Oak Ridge, Hollaender established a new symposium series modeled on the style of those convened at Cold Spring Harbor. In his obituary to Alex, who died in 1986, Jack von Borstel, then at the University of Alberta, Canada, related that in 1954 Hollaender was planning to bring Linus Pauling to Oak Ridge as a speaker in his new symposium series. When the then chairman of the Atomic Energy Commission, Admiral Lewis Strauss, got wind of this he was greatly concerned that Pauling's presence at a National Laboratory might fuel another of Senator Joe McCarthy's investigations into Pauling's suspected communist leanings. So he promised Hollaender additional funds for the symposium if he would hold it anywhere but Oak Ridge. Alex moved the meeting to nearby Gatlinburg, Tennessee which became the permanent home of the highly successful and internationally recognized Gatlinburg Symposia.

The Hollaender years at Oak Ridge were vintage in the history of that Laboratory, fondly and proudly recalled by its many distinguished staff. Jack von Borstel recalls that the Sunday walks in the Cumberland Mountains led by Alex and in later years by Dick Setlow, were "joyful times spent fossil hunting, discussing the future of art and science, and planning new ventures.... Visitors from around the world were invited on these walks, and fossil stigmata of *Lepidodendron* from the Cumberlands now grace the desks and tables of biologists from Europe, Asia, and North and South America."

Hollaender's presentation at the 1941 Cold Spring Harbor Symposium was entitled "Wavelength Dependence of Mutation Production in the Ultraviolet with Special Emphasis on Fungi." His paper documented studies on the effects of UV radiation on microorganisms that he and his associates had initiated in the early nineteen-thirties. As early as 1928, Frederick Gates of the Rockefeller Institute had noted that the wavelengths of light most effective for killing bacteria were those that were most efficiently absorbed by nucleic acids. Gates prophetically commented that this observation "has a wider significance in pointing to these substances [nucleic acids] as essential elements in growth and reproduction."

In a paper published in the proceedings of the 9th International Congress on Photobiology held in Philadelphia in 1984, John Jagger noted that Gates was really more ambivalent about the action spectrum of UV radiation. In fact, Gates concluded that the action spectrum for killing of *Staphylococcus aureus* and its phages resembled curves "for the specific absorption of ultraviolet light by protoplasm, by proteins, by certain amino acids and nucleoproteins, and by certain enzymes." As Jagger succinctly pointed out, "clearly Gates was leaving his options open!"

Hollaender also recognized that since different biological molecules optimally absorb UV light of different wavelengths it might be possible to modify different metabolic activities in cells by exposing them to selected wavelengths. This emphasis led him to the extensive use of monochromatic UV radiation, that is, radiation of a single wavelength. It is sobering to reflect that at the 1941 meeting, presumably at the very time that Avery and his colleagues were meticulously tracking the chemical nature of the active principle in pneumococcal transformation, Hollaender and Emmons reported that the mutagenic profile of different wavelengths of UV radiation (referred to in the vernacular as the UV action spectrum) was remarkably similar to the absorption profile of different wavelengths of UV radiation by nucleic acids (the absorption spectrum). This observation is pregnant with the implication that the target for UV radiation, that is, the gene, is composed of nucleic acid. Indeed, this implication was specifically documented by Hollaender and Emmons.

> We have tested intensively the effectiveness of eight wavelengths between 2180 and 2967 Å in their ability to produce mutations. There are several interesting features in these [mutational and fungicidal] curves. First, 2650 Å appears to be the wavelength most effective in producing mutations as well as toxic action. . . . The 2650 Å maximum coincides with the high absorption coefficient of nucleic acids near this wavelength.

However, reflecting reservations very similar to those expressed by Avery and his colleagues three years later, and possessed with far less persuasive experimental evidence, Hollaender and Emmons appropriately concluded that:

> This does not necessarily mean that nucleic acid is the only cell component responsible for this maximum. Proteins and certain enzymes which are present in only very low concentrations could contribute very well to the [absorption] maximum at [2650 Å].

The concluding paragraph of Hollaender and Emmons's paper instructs us once again how deeply ingrained the prevailing dogma of the protein structure of the gene in the pre-Watson and Crick era really was.

> It is probably somewhat dangerous to overemphasize the importance of nucleic acid in the study of radiation effects on living cells. It is very well possible that in radiation-produced mutations, the nucleic acid is only the "absorbent" agent.

This notion was reinforced by Muller who commented on Hollaender's results in his summary of the symposium.

> The type of dependence of the mutation frequency upon the wave length of the [UV radiation] found by Hollaender is of special interest. Not only does the curve show the same peak as that shown by the ultraviolet absorption of nucleic acid, ... but in addition it shows a secondary peak, at shorter wave lengths, as expected for the protein absorption. This result would lead to the conclusion that ultraviolet absorption by either component of the nucleoprotein is capable of leading to mutation, and that accordingly, in one or the other of the cases at least, there is a transfer of the energy, or a chain of reactions, starting at the absorbing group, and secondarily reaching the material in which the final mutational change occurs.

The spectroscopic correlation in support of the chemical nature of the gene provided by Hollaender and his coworkers was much less direct than the biochemical evidence published shortly thereafter by Avery and his coworkers. But even if the general scientific community in the early nineteen-forties had been more quickly receptive to the notion that genes were composed exclusively of DNA, it was quite something else to prove that DNA had a genetic, rather than a purely structural, function. Judson recounts a discussion with Max Delbrück at Cold Spring Harbor in 1972, during which Delbrück retrospectively debunked the notion of Avery's reticence and uncertainties about the implications of his experimental results. On the contrary, Delbrück was of the opinion that the budding community of phage geneticists (the famous "phage group") was well aware of Avery's work, and he of theirs. The problem according to Delbrück was that everybody "who thought about it was confronted with this paradox, that on the one hand you seemed to obtain a specific effect [bacterial transformation] with DNA, and on the other hand at that time it was believed that DNA was a *stupid* substance, a tetranucleotide which couldn't do anything specific."

Delbrück related to Judson that even when the structure of DNA was revealed it was not immediately obvious "how you use this Watson-Crick structure to code for anything to carry specificity." Much of the revolution that characterized the 15-year period following Watson and Crick's seminal hypothesis on the structure of DNA concerned the systematic marshaling of incontrovertible evidence that DNA indeed embodies a genetic code, and how this code is deciphered.

A final word on the much discussed question of Avery, MacLeod, and McCarty's discovery and its skeptical reception in the general scientific community, if not among the small but influential group of phage geneticists. Prompted by Joshua Lederberg, who had just accepted the presidency of Rockefeller University, and who from the outset was greatly impressed by their work, Maclyn McCarty wrote his own accounting of this celebrated period in the history of genetics in 1985. McCarty called his book *The Transforming Principle: Discovering that Genes Are Made of DNA*. In the final chapter, entitled "Aftermath," he enumerated several events that in his words "certainly influenced the reception of the pneumococcal DNA story." He cited the fact that

1944 was the height of World War II and was not noted for the most avid attention to the scientific literature, particularly by biomedical scientists in Europe. He also noted that "the *Journal of Experimental Medicine* was not read by many geneticists and general biologists," if at all, and emphasized that in the early nineteen-forties bacteria were not yet considered appropriate experimental organisms for genetic studies.

> The majority of geneticists were not prepared to accept information emerging from studies of the pneumococcus as having any bearing on the genetics of higher organisms, just as the majority of biochemists were still influenced by the notion that nucleic acids were monotonously alike and therefore not likely candidates for the possession of biological specificity.

At this time in the history of genetics, the use of microorganisms for genetic studies was indeed considered something of an anathema. Judson wrote that:

> Conventional bacteriologists were by no means clear that microorganisms could be thought about genetically or in terms of natural selection. Many believed that resistance [to antibiotics or to phage infection] was some kind of adaptation induced, in a few of the bacteria in a culture, by the exposure to the antibacterial agent; the mechanism of induction was never specified, but one might imagine a metabolic change so drastic that it would rarely succeed but when it did would be passed on to the progeny. The idea smacks of the pre-Mendelian, pre-Darwinian notion of the inheritance of acquired characteristics; Luria damned bacteriology as "the last stronghold of Lamarckism."

It is thus quite remarkable that in addition to Avery and his colleagues and Delbrück and Luria (whose bold foresight in their adoption of phage as genetic tools has been extensively acknowledged), Alex Hollaender was also considerably persuaded by the advantages of both UV radiation and microorganisms over the more cumbersome and considerably more expensive requirements of an X-ray source and the use of higher eukaryotes such as *Drosophila* for mutational studies. Finally, and not without a trace of bitterness, MacLeod recounted in some detail the staunch recalcitrance and prolonged negative lobbying of Alfred Mirsky, a biochemist at the Rockefeller, who according to Judson was "trying to prove that the protein associated with nucleic acids in the chromosomes of higher organisms was the active component."

Aside from the notion that they were made of proteins, other dogmas about the nature of genes in the nineteen-thirties and nineteen-forties hindered informed opinion about gene damage and its repair. Robert H. Haynes, now Emeritus Professor at York University in Ontario, Canada, has long been a devoted historian of genetics in general and the field of DNA repair in particular. Later chapters will recount Haynes's personal contributions to the discovery and elucidation of excision repair of DNA. In the spring of 1995, Bob Haynes and I renewed our long acquaintance at an international conference on DNA repair and mutagenesis in the tranquil splendor of Taos, New Mexico, and enjoyed an extended discussion about the historical climate in which the DNA repair field struggled to emerge. The time was an auspicious one, historically. The Taos meeting celebrated the anniversary of a meeting that Haynes helped organize exactly 30 years earlier; a meeting widely acknowledged as the first formal conference dedicated to the topic of DNA repair. That meeting, held in Chicago in 1965, was attended by a little over 40 conferees. The Taos meeting convened 30 years later was attended by more than tenfold that number of scientists. The Taos meeting also celebrated the 60th anniversary of Alex Hollaender's published annotation of post-UV radiation recovery in bacteria, a milestone in the history of the field to which I will return later in the chapter.

Haynes sensitized me to several important nuances that in his view influenced the delayed emergence of the field of DNA repair. He pointed out that, contrary to contemporary knowledge about the intrinsic instability of DNA as a biological macromolecule, the highly prevalent sentiment in the decades of the nineteen-thirties and nineteen-forties was precisely the opposite. Whatever their true chemical composition, genes were conceptualized as incredibly stable elements. "I think it's fair to say," Haynes remarked, "that in the early decades of this century cells were regarded by many people as black boxes and genes were regarded as black beads in the black boxes. It was widely thought that genes were physically stable; that they were somehow sequestered in cells; that they were immune to attack by environmental agents; that they replicated *precisely*—I wouldn't use the term *accurately*, because the notion of replication of genetic information had not yet come forward. Another point is that physiological conditions inside the cell were considered to be warm and friendly and safe for genes. The existence of a potentially substantial mutagenic burden on living cells was completely unrecognized because chemical mutagens were not known, and the doses of ionizing radiation in normal environments were very low. Similarly, the kind of UV radiation that produced mutations experimentally was not present at high levels in normal environments."

As early as 1909, the Danish biologist Wilhelm Johannsen had described mutations as rare, sudden, discrete events that caused genes to change from one stable state to another. As Haynes put it, "very early on, the Mendelian atomistic concept of the gene was based on the notion of inherent stability." Indeed, the choice of X rays by Muller and by Stadler as an agent with which to produce mutations in *Drosophila* and in maize, respectively, was based on the notion that genes would be too stable to modification with anything less energetic.

This view was reinforced by the writings of Max Delbrück, and later by the renowned physicist Erwin Schrödinger. Delbrück, one of the earliest and perhaps the most celebrated of the group of physicists turned biologists in the nineteen-thirties and nineteen-forties, is of course best known for his contributions to the ascendancy of molecular genetics and molecular biology in the decades of the nineteen-forties and the nineteen-fifties. Much of his life's work is discussed in Judson's historical discourse on the biological revolution that began in the nineteen-fifties, and in the more recent formal biography of Delbrück by Ernst Peter Fischer and Carol Lipson, entitled *Thinking About Science: Max Delbrück and the Origins of Molecular Biology*. Delbrück did his graduate work at an intellectual Mecca of physics in the mid nineteen-twenties, the Department of Physics at the University of Göttingen, which included among its faculty Max Born, Wolfgang Pauli, and Werner Heisenberg. Following a brief stay in Bristol, England, Delbrück accepted a postdoctoral fellowship with

Niels Bohr in Copenhagen. Bohr proved to be highly influential in Delbrück's future thinking. In particular, a lecture by Bohr delivered in Copenhagen in August 1932 is credited with changing the course of Delbrück's scientific career. In this lecture entitled "Light and Life," delivered at the International Congress of Light Therapists, Bohr suggested that:

> There might be a complementary relation between life and atomic physics analogous to the complementarity encountered with the wave and particle aspects in atomic physics. The result would be a kind of uncertainty principle concerning life, analogous to Werner Heisenberg's uncertainty principle in quantum mechanics.

This notion of "complementarity" intrigued Delbrück enormously. During the preparation of his intended autobiography, which was tragically interrupted by his terminal illness, Delbrück wrote that:

> In physics it is obvious that even in the simplest case such as a proton running around an electron one can do classical physics until one's dying day and never get a hydrogen out of it. In order to achieve this, one has to use the complementarity approach. If one looks at even the simplest kind of cell, one knows it consists of the usual elements of organic chemistry and otherwise obeys the laws of physics. One can analyze any number of compounds in it but one will never get a living bacterium out of it, unless one introduces totally new and complementary points of view.

Delbrück began his search for complementarity in biology by exploring the effects of ionizing radiation on genes. In 1932, Hermann Muller visited Berlin to extend his studies on comparative mutagenesis by X rays with the Russian émigré Nicolai Timoféeff-Ressovsky, director of the Genetics Laboratory at the Kaiser Wilhelm Institute for Brain Research in Berlin-Buch. Fischer and Lipson tell us that at about this time Delbrück initiated a discussion group at his mother's house. K. G. Zimmer, who attended many of these discussions commented that it was commonplace to "talk for 10 hours or more without any break, taking some food during the sessions." Soon after Muller left Berlin, Delbrück began an active collaboration with Timoféeff-Ressovsky and Zimmer that culminated in a quantum model of gene mutation.

The quantum model of gene mutation elaborated by Delbrück, Timoféeff-Ressovsky, and Zimmer was published in 1935 under the title "About the Nature of the Gene Mutation and the Gene Structure." It turned out to be incorrect of course, but it was "a successful failure" because it provided an important stimulus to modern biology and constitutes one of the earliest approaches to the discipline that we now formally designate as molecular biology. Delbrück and his colleagues argued that genes may be composed of extraordinarily stable molecules, perhaps constituting some hitherto unrecognized state of matter. His notion of the gene was purely physical, and this mind-set readily lent itself to the sentiment that spontaneous mutations could only arise from quantum-statistical fluctuations in the isomeric states of the genetic molecules, hence their extreme rarity. Similarly, the changes produced by X rays that so dramatically increased the yield of mutations were considered to be undefined quantum events in genes; events that might be expected to result only from high energy perturbations produced by agents such as X rays. As Evelyn Witkin succinctly put it many years later in her historical reflections about mutagenesis, "the prevailing notion until the late nineteen-forties was that mutations were instantaneous events—the mutagen went 'Zap!' and that was that."

Delbrück's paper was published in German and was presumably not widely read in the United States. Nonetheless, his conception about the nature of the gene inspired Schrödinger to publish a small but highly influential book in 1944 entitled *What Is Life?*, based on a series of public lectures that he delivered at Trinity College, Dublin a year earlier. In his book, Schrödinger expounded on and popularized many of Delbrück's ideas about the structure of the gene and the nature of mutational change. Indeed, Schrödinger devoted an entire chapter of his book to Delbrück's physical theory of genetic stability and change. He stated that:

> We may safely assert that there is no alternative to the molecular explanation of the hereditary substance. The physical aspect leaves no other possibility to account for its permanence. If the Delbrück picture should fail, we would have to give up further attempts.

In a recent analysis of *What Is Life?*, coincident with its 50th anniversary, G. Rickey Welch tells us that the fundamental paradox that Schrödinger wished to address was how a *single* macromolecule, the gene, could determine dynamically "the very orderly and lawful events within a living organism," whereas most physical laws deal with systems entailing *very large* numbers of particles. His resolution of this paradox led him to the suggestion that the gene was composed of a thermostable macromolecular "aperiodic crystal" that was held together by quantum forces and embodied a "miniature code." The predominant message of *What Is Life?*, and the one that most passionately fired the imagination of many who read the book, was this notion of a code. But the concept of genetic stability was simultaneously, albeit secondarily, reinforced. Schrödinger's book was widely read, especially by biophysicists. It apparently had a profound influence on a young Jim Watson, who encountered it as an undergraduate at the University of Chicago soon after it was published. Judson informs us that in a retrospective essay about his graduate student days, contributed to a Festschrift on the occasion of Max Delbrück's 60th birthday, Watson wrote that after reading Schrödinger's book he "became polarized towards finding out the secret of the gene." In more recent autobiographical reflections, Watson recalled that:

> I was 17, almost three years into college, and after a summer in the North Woods, came back to the University of Chicago and spotted the tiny book *What Is Life* by the theoretical physicist Erwin Schrödinger. In that little gem, Schrödinger said that the essence of life was the gene. Up until then, I was interested in birds. But then I thought, well, if the gene is the essence of life, I want to know more about it.

By the mid nineteen-thirties, Delbrück had become deeply immersed in biology. As already indicated, Warren Weaver of the Rockefeller Foundation (of which Delbrück was a fellow) was extremely interested in promoting the integration of physics and biology and helped support the first Physico-Biological Conference in Copenhagen in 1936, attended by Bohr, Delbrück, and Timoféeff-Ressovsky. The Foundation also established a program called the *Special Research Aid Fund for Deposed Scholars*, which was specifically designed to assist the relocation of eminent and promising scientists in Europe who wished to escape the Nazi regime. When Delbrück was approached about assistance through this program, he initially elected to go to England to work with R. A. Fisher, J. B. S. Haldane, and Cyril Darlington on the theory of natural selection. Providentially for molecular genetics, Haldane was planning to be away from his laboratory for an extended period, so Delbrück proposed studying genetics in the United States instead. In late 1936, he wrote to the Rockefeller Foundation that:

> I thought that if I was to pursue work on mutations further, a stay in the United States where all the main work on mutations is being done would be more profitable. In particular, it would of great help to me, if I could stay for some while in Pasadena in order to learn from T. H. Morgan and his co-workers, then in Chicago, to work and discuss Natural Selection with Sewall Wright, whose work I value very highly, and finally in Cold Spring Harbor, to discuss with Demerec his work on mutable genes, which is most interesting to Timoféeff and myself.

In 1937, Delbrück left Nazi Germany for the West. Parenthetically, his friend and colleague Timoféeff-Ressovsky returned to his native Russia where he was systematically accused of collaborating with the Nazis. Additionally, he was displaced from mainstream genetics by the ascendance of Lysenkoism. He lived out the remainder of his life in Russia in isolation and intellectual restriction and died in 1981, the same year as Delbrück. Delbrück's beginnings in the United States in 1937 were as a Rockefeller Fellow at Caltech, "intent on discovering how his background in physical sciences could be productively applied to biological problems." He was introduced to the mysteries of bacteriophage and their reproduction in bacteria by Emory Ellis and became deeply fascinated by the many questions raised by the study of this simple biological system. His introduction to a lecture that Delbrück delivered to the Harvey Society in New York in 1946 is most illuminating in this regard.

> You might wonder how such naïve outsiders get to know about the existence of bacterial viruses. Quite by accident, I assure you. Let me illustrate by reference to an imaginary theoretical physicist, who knew little about biology in general, and nothing about bacterial viruses in particular, and who accidentally was brought into contact with this field. Let us assume that this imaginary physicist was a student of Niels Bohr, a teacher deeply familiar with the fundamental problems of biology, through tradition, as it were, he being the son of a distinguished physiologist, Christian Bohr.
>
> Suppose now that our imaginary physicist, the student of Niels Bohr, is shown an experiment in which a virus particle enters a bacterial cell and 20 minutes later the bacterial cell is lysed and 100 virus particles are liberated. He will say: "How come, one particle has become 100 particles of the same kind in 20 minutes? That is very interesting. Let us find out how it happens! How does the particle get in to the bacterium? How does it multiply? Does it multiply like a bacterium, growing and dividing, or does it multiply by an entirely different mechanism? Does it have to be inside the bacterium to do this multiplying, or can we squash the bacterium and have the multiplication go on as before? Is this multiplying a trick of organic chemistry which the organic elements chemists have not yet discovered? Let us find out. This is so simple a phenomenon that the answers cannot be hard to find. In a few months we will know. All we have to do is to study how conditions will influence the multiplication. We will do a few experiments at different temperatures, in different media, with different viruses, and we will know. Perhaps we may have to break into the bacteria at intermediate stages between infection and lysis. Anyhow, the experiments only take a few hours each, so the whole problem can not take long to solve."

Perhaps you would like to see this childish young man after eight years, and ask him, just offhand, whether he has solved the riddle of life yet? This will embarrass him, as he has not got anywhere in solving the problem he set out to solve. But being quick to rationalize his failure, this is what he may answer, if he is pressed for an answer: "Well, I made a slight mistake. I could not do it in a few months. Perhaps it will take a few decades, and perhaps it will take the help of a few dozen other people. But listen to what I have found, perhaps you will be interested to join me."

Following a wartime sojourn at Vanderbilt University in Nashville, Tennessee, Delbrück returned to Caltech with an established phage group. Those in the DNA repair field who knew him then were uniformly and indelibly impressed. Bernard S. (Bernie) Strauss, Professor of Molecular Biology at the University of Chicago, promoted several important areas of DNA repair during the course of his career, which we will consider later. Strauss did his graduate work in the Biology Division at Caltech and remembers Delbrück well. "The Biology Department was small enough at the time so that we got to know the work of the other groups," remarked Strauss. "Max Delbrück had just come to Caltech and with him the phage group. There was no way one could be a graduate student at Caltech and not be influenced by Delbrück. His bias, partly derived from that of Bohr, was that there would be shown to be some indeterminacy principle in biology by which the laws of physics and chemistry would be seen to be inadequate for organisms. He never found them. Partly because of his background in physics, Delbrück was not particularly interested in metabolic pathways, or for that matter in most of the biochemical studies of the time. The phage experiments were nonchemical, as was most of genetics before Beadle's revolution."

Delbrück's demeanor at scientific seminars is legendary. Strauss related that "he would interrupt even the most distinguished speaker time and time again with the comment 'I don't understand you.' This comment, repeated often enough, drove strong men to tears and we graduate students waited with bated breath to see when it would happen and how it would affect the speaker. I eventually realized that these interruptions could mean one of two things. The first was that he truly didn't understand something and wished clarification. Most of us, I think, are hesitant enough that we tend to remain silent rather than risk appearing foolish in public when we miss something at a seminar. Not so Delbrück. He supposed (or acted as if he supposed) that he was smart enough that if *he* didn't understand something, the matter hadn't been clarified sufficiently. A more insidious reason for his interruptions was that he had detected a fault in the speaker's logic and then 'I don't understand you' was a challenge to justify oneself." Strauss was of the view, one shared by many, that Delbrück was able to impose his will and intellect on a number of scientists, many of whom were major intellectual figures in their own right. "But he also made mistakes," Strauss pointed out. "He completely ignored the evidence for lysogeny for a long time (an issue we will revisit in a later chapter). And he and his group were particularly disparaging of the early biochemical work of Seymour Cohen on bacteriophages."

Delbrück was a member of Strauss's Ph.D. thesis committee and Strauss was petrified in anticipation of the terror that Delbrück could inflict on a graduate student during the question and answer period. "However," recalls Strauss, "my thesis was on vitamin B6 metabolism in pH-sensitive mutants of

Neurospora. This topic held little interest for him and he disdainfully announced to me ahead of time that he was not even going to read my thesis. I confess to a tremendous sense of relief at hearing that!"

"I can't help but view Delbrück as a sort of superhuman intellect different from any other I encountered in my student career at Caltech," Strauss concluded. "Beadle and Horowitz were terrific scientists, but Delbrück was different. He not only drove the phage group, but his insights on the *Neurospora* work that was going on at Caltech at that time, were central. He was a tremendous teacher and ran a phage course for a small group of Caltech graduate students. We were absolutely fascinated with this course. One time he was out of town and was replaced by an unknown (to us) mild-mannered visitor who mumbled, was extremely difficult to understand, and in general confused us a lot. We later learned that his name was Al Hershey!"

Delbrück was an individual with far-ranging vision and eclectic scientific tastes and, aside from his primary interest in phage genetics, he impinged on the field of DNA repair and mutagenesis in more than a casual way, as later chapters will attest. Returning to the notion of genetic stability that he so convincingly promulgated, a reconsideration of this notion might reasonably have been prompted by the demonstration that in addition to X rays, certain chemicals are also potent mutagens. In 1938, Hermann Muller visited the Institute of Animal Genetics at the University of Edinburgh, where he met Charlotte Auerbach and inspired her to investigate the nature and causation of radiation-induced mutations in *Drosophila*. The similarities between the known pharmacological effects of X rays and the alkylating agent mustard gas led Auerbach to pursue studies with mustard gas instead, and to the immediate discovery of a clear mutagenic effect by this chemical. But the vicissitudes of world politics, specifically the concern that mustard gas might be used against the British during World War II, preempted the general announcement of this important finding for a number of years. At the 1951 Cold Spring Harbor Symposium on *Genes and Mutations*, Auerbach recalled that:

> It is just ten years ago that Dr. Muller . . . , at the last Cold Spring Harbor Symposium on the gene and mutation, said that attempts to induce mutations by chemical means had not yet given any clear positive results. As a matter of fact, this statement was then no longer true. Clear positive results had been obtained only a few weeks earlier with mustard gas . . . , but owing to a security ban on publication they had to be kept secret until some time after the war . . .

Charlotte Auerbach Lotte died in 1994 at the age of 95. In a posthumous tribute, her colleague at the University of Edinburgh for many years, B. J. Kilby, commented that:

> This delay had its advantages because Lotte was able to conduct an unhurried and careful comparison of the genetic effects of alkylating agents and ionizing radiation. By the time publication was permitted she had an impressive corpus of observations to present to the scientific community . . .

Bob Haynes is of the viewpoint that the intellectual predilection that genes were intrinsically stable entities was recapitulated even after it was clearly established that they were comprised of DNA. He recounted a visit in the early nineteen-sixties with the cell biologist Daniel Mazia, "the guru of mitosis." At that time, Haynes and others were actively speculating about the existence of a DNA repair mode by which damaged nucleotides were excised (presumably enzymatically) from the genome of bacterial cells. Mazia debunked the idea of excision repair as being inconsistent with the well-documented "fact" that DNA did not turn over. In a scholarly review entitled "Physiology of the Cell Nucleus" published in 1952, Mazia wrote that "to the observer who is impressed with the 'dynamic state' of cell constituents generally, the stability of DNA is remarkably in accord with what might be expected of genetic material." In fairness to Mazia, I wish to stress that this statement is quoted out of his intended context to emphasize the prevailing belief that DNA was a stable molecule, a sentiment that was antithetical to the idea that it could serve as a substrate for degradative reactions.

Franklin Stahl of the famous Meselson-Stahl DNA replication experiment, some of the history of which is recounted in a later chapter, was very much at the center of the emergence of the discipline of molecular biology in the early nineteen-fifties. When I challenged him with the question as to why the concept of gene/DNA repair was late in coming, he responded with a brief but illuminating reply:

> I suspect because of a widespread belief (unspoken I suspect, but amounting to worship) among geneticists that *the genes* are so precious that they must (*somehow*) be protected from biochemical insult, perhaps by being carefully wrapped. The possibility that *the genes* were dynamically stable, subject to the hurly-burly of both insult and clumsy (i.e., enzymatic) efforts to reverse the insults, was unthinkable.

Regardless of the many subtle but interacting contributions to the conceptual poverty about the inherent *instability* of DNA as we know it today, its increasingly obvious vulnerability to mutational alteration by physical (and chemical) agents did provoke a trickle of provocative allusions in the late nineteen-thirties and the nineteen-forties to phenomena that we now consider the province of cellular responses to genomic injury, that is, DNA repair. As early as 1920, G. A. Nadson documented the recovery of some cells in a population of yeast following radiation sickness induced by their exposure to ionizing radiation. The notion of "radiation sickness" in microorganisms carried over into the literature of the early nineteen-fifties, when Richard B. Roberts and his colleague Elaine Aldous borrowed this phrase to describe the recovery of viability of *E. coli* from exposure to UV radiation.

In 1918, William T. Bovie, then at Harvard Medical School, ventured the opinion that none of the changes observed in UV-irradiated cells are initial changes, but represent aftereffects that vary as a function of metabolic processes in the cell. According to Paul Swenson, a longtime member of the scientific staff at the Oak Ridge National Laboratory, who thoroughly documented (but never published) Bovie's contributions to early photobiology, Bovie attempted to localize the action of UV radiation by employing two important principles; the selective absorption of the radiation and the noted hypersensitivity of cells to the radiation. During his studies on *Paramecium caudatum* using monochromatic UV radiation at 280 nm, Bovie noted division delay before resumption of cell division. Among other possible explanations that he offered to explain this phenomenon was "the inability of the nucleus to undergo rapid *recovery* from the injury caused by the radiations." Despite limited experimental data, Bovie prophetically inferred that the primary site of action of UV radiation, both in the production of photochemical damage and in the recovery from such damage, was the nucleus. In a draft of an aborted history of DNA repair written in the late nineteen-sixties, Swenson commented that:

> The road to our present-day understanding of UV and X-ray damage to cells and the repair of this damage has been one with numerous detours and temporarily impossible barriers. Many roads have been followed into regions that have left the biologist bewildered and confused. Bovie recognized that for full understanding of the radiation problem the chemical make up and the physiology of the cell must be understood.

Swenson quoted Bovie as stating that:

> Skilled in the art of using radiation, we possess a new tool with unique and invaluable possibilities for scientific investigation. The results of our investigations will be contributions not only to the nature of action of the radiation but also to the nature of life processes.

This cursory sweep through the literature of the early 20th century documents but a few of (presumably) many other provocative allusions to the effects of radiation on the genetic material of cells and their potential for recovery from such effects.

In 1935, Alex Hollaender was in charge of radiation work for the National Research Council Project in Wisconsin at his alma mater, the University of Wisconsin. He and his associate John Curtis noted that if *E. coli* cells were exposed to UV radiation and then plated on nutrient agar, surviving colonies appeared later than did colonies of bacteria shielded from the radiation. In a discussion of this growth lag in a brief paper for the *Proceedings of the Society for Experimental Medicine*, Hollaender tersely noted that "the possibility of recovery of the irradiated bacteria is not entirely excluded." This little paper is widely regarded as perhaps the earliest, albeit indirect, reference to DNA repair in the literature, prompting Phil Hanawalt and Dick Setlow to reprint it under the title "An Early Suggestion of DNA Repair" in the proceedings of a symposium on DNA repair held in Squaw Valley, California in early 1974. Hollaender and Emmons extended their bacterial studies to fungal spores and once again noted that if the irradiated spores were held in a simple salt solution devoid of the nutrients required for normal growth instead of being plated out on growth medium immediately, "there was a marked tendency to recover." They also observed variations in the yield of UV radiation-induced mutations as a function of these postirradiation treatments. This is what Hollaender had to say about these early observations 40 years later in a foreword to the Squaw Valley symposium proceedings.

> An "age" has passed in the 40 years since we first observed recovery from radiation damage in irradiated bacteria. During the early nineteen-thirties, we had been discussing the possibility of rapid changes after radiation exposure with Farrington Daniels, Benjamin Duggar, John Curtis, and others at the University of Wisconsin. After working with living cells, we had concluded that organisms receiving massive insults must have a wide variety of repair mechanisms available for restoration of at least some of the essential properties of the cell. The problem was how to find and identify these recovery phenomena. I realized there must be some enzyme that stimulated recovery, and I spent a few weeks in [John Howard] Northrop's laboratory (then called Rockefeller Institute) outside Princeton, New Jersey, working with Kunitz in an effort to isolate the stimulating enzyme. But with my limited

knowledge of the recovery process our experiments were not successful. In the nineteen-forties, we attempted to interest other investigators in the problem of recovery, but to no avail.

Other research groups documented postirradiation recovery phenomena after exposing microorganisms to UV light. Postirradiation recovery when irradiated bacterial cells were stored in the cold was reported by the French radiobiologist Raymond Latarjet in 1943. Recovery of UV-irradiated *E. coli* was also reported by Roberts and Aldous in 1949. These investigators explored a number of variables associated with growth prior to exposure to UV radiation, the conditions associated with the irradiation itself, and the postirradiation conditions. They observed significant levels of recovery under optimal conditions, and these were accompanied by marked changes in the shapes of the survival curves. They concluded that:

> The primary effect in the killing . . . by ultraviolet is the production of a poison within the cell. Under certain conditions this poison can be inactivated or removed and the cell remains viable.

The term "poison" with respect to the effect of UV radiation on microorganisms surfaced in the literature quite frequently at this time. The term was never used explicitly to connote the formation of a chemical poison and one gains the impression that some authors wished to imply "poisoning" of the gene in a metaphorical sense. An example of this metaphorical use of the term is provided in the next chapter on the discovery of photoreactivation. However, no one stated this analogy formally and hence no author was prompted to the congruent inference of the *repair* of "poisoned" genes as the basis for postirradiation recovery.

As already mentioned, Salvador Luria together with Max Delbrück founded the famous "phage group" which revolutionized molecular genetics in the nineteen-forties and nineteen-fifties. Their contributions to molecular biology are now legion, in particular their classical fluctuation analysis which showed that mutations in *E. coli* that conferred a particular phenotype (such as antibiotic resistance) arose spontaneously and were not specifically induced by the selective agent itself. This conclusion might sound trifling in 1996, but as alluded to earlier, in 1943 it was by no means obvious that microorganisms experienced "real" spontaneous mutagenesis. In fact, the contrary position was widely held, and this was one of the principal reasons that higher organisms were preferred for genetic studies. The inspiration for the celebrated fluctuation experiment that led Luria and Delbrück to the conclusion that mutations were stochastic events in bacteria, apparently stems from Luria's observation of people playing with a slot machine at a party. In his autobiography published in 1984 entitled *A Slot Machine, A Broken Test Tube*, Luria tells us that while watching the slot machine he was led to the abrupt insight that if he could divide a bacterial culture into multiple very small cultures, each of which ideally comprised just a single cell, mutations that arose spontaneously very early in the history of any single culture should accumulate in that population, and hence should "pay out" large jackpots (just as slot machines occasionally did) of mutants. In contrast, the cultures in which there were no spontaneous mutants, or only a few because they arose late in the history of the culture, should not yield such handsome dividends. These fluctuations in mutation frequency were indeed observed experimentally and quantitatively fulfilled the predictions of a mathematical model contributed by Max Delbrück, thus proving that mutations arise spontaneously in bacteria.

A lesser known aspect of Luria's career is that he, sometimes with his graduate student Renato Dulbecco, made a number of experimental observations that impinged directly on several aspects of cellular responses to DNA damage. In his autobiography, Luria traced the tortuous course that led him, like Hollaender, to the persuasion of the enormous utility of microorganisms and radiation for the study of the gene. He began his formal training as a physician at the University of Turin. Like many molecular biologists since, Luria did not relish the prospect of devoting his career to the care of the sick. He wrote that he:

> Never fully identified with [medicine] as an activity. I was perhaps too immature to integrate myself into the confusing structure of the hospital, and lacked the initiative, even eagerness, required in an Italian medical school to gain access to practical expertise beyond the formal lectures. Thus, although I graduated in 1935 near the top of my class I felt totally inadequate as a doctor and not at all eager to practice.

Luria was greatly influenced at that time by Ugo Fano, a friend and fellow student at the University of Turin who was studying physics with Enrico Fermi and who inflamed Luria's growing, and by his own admission, "romantic view of physics" by recounting to him the startling revolution in that discipline in the nineteen-twenties and nineteen-thirties. Fano went on to evolve a highly distinguished career in nuclear medicine and recently received the Enrico Fermi Award, granted by the U.S. government in recognition of achievements in the field of nuclear energy. Luria decided to "find a medical specialty that could bring me close to physics and then use it as a bridge to what might be called 'biophysics.'" Upon graduating from medical school, he joined a radiology department in Turin and registered for a course there. Luria wrote that his:

> Disappointment and sense of loss were profound. Both radiotherapy and diagnostic radiology turned out to be the dullest of medical specialties . . . And the so-called specialization courses turned out to be a farce: the chief professor could not write the simplest equation correctly even when copying from a manual. And the scheduled physics course for radiologists was not taught at all.

Even in this period that preceded the discovery of the structure of DNA and the full flowering of molecular biology, many aspiring young scientists curious about the gene and its function were disappointed to find that central questions in genetics were not part of the investigative lexicon of classical radiobiology. As mentioned earlier, the perception is widely held that although radiobiology was a discipline that emerged from the exploration of biological responses to radiation damage, it lacked the vibrancy and dynamic curiosity, as well as the experimental excitement and rigor of molecular biology. This indictment persisted well beyond the time that molecular biology became an established discipline. Nor was it totally undeserved. For while molecular biologists steeped themselves in fundamental questions of central interest and devised cunning experiments with informative model systems and innovative technologies to answer them, many radiobiologists were content to repeat survival curve after survival curve, apparently persuaded that detailed analysis of variations in the shape of such curves would yield secrets about the cellular target for radiation injuries, the nature of the injuries, and how cells responded to them. As already mentioned, a regrettable outcome of this viewpoint was that radiobiological phenomena, including DNA repair, were considered in many quarters to be the province of mediocre scientists exploring mediocre problems. Of course the definition of mediocrity is often in the eye of the beholder. The intellectual centerpiece of the golden age of molecular biology comprised a small group of elitists often referred to as "the inner circle," who had a very restricted view of "good" and "mediocre" science. Matthew Meselson, a fully-fledged member of this hallowed group, who with Franklin Stahl performed the famous experiment that now bears their names, admitted to Horace Judson that even he experienced the scorn of the phage group when he decided to work with bacteria.

> Doing this [DNA replication] experiment with *E. coli* was a special kind of leap, in that very sophisticated intelligent people worked with phage T4, but a rather dull lot of people worked with bacteria—or maybe, people who weren't liked so much in our lab. These things were never spoken but they float around at the back of your mind somewhere. But Caltech was Delbrück and the phage group, and people who worked on bacteria were—shall we say, outside the circle.

Luria's dim view of radiobiology notwithstanding, when he eventually gravitated to phage genetics, his background in and understanding of radiobiology served him well through several contributions to the DNA repair field. In what he later described as a critical turning point in his career, Luria left Turin for the University of Rome where he finished his courses in radiology and studied physics for a year in the department where Enrico Fermi was working. During that year, one of the physicists in Rome introduced him to the German writings on the gene by Max Delbrück. Luria's response to this exposure bears an uncanny parallel to that of his protégé, Jim Watson, to Schrödinger's writings in later years.

> These papers seemed to me, however ignorant of genetics I then was, to open the way to the Holy Grail of biophysics—and there and then, I believe, I swore to myself that I would be a knight of that Grail.

Luria recounts that he was profoundly influenced by Delbrück's notion of the gene as a molecule based on his interpretation of Muller's studies showing the mutagenic action of X rays in *Drosophila*. Luria recognized that higher eukaryotes such as *Drosophila* were not well suited to examine Delbrück's ideas. Homozygous mutations in diploid cells are too rare, even after exposure to X rays, and thus one would have to work with a very large number of organisms to obtain meaningful data. When he read Delbrück's papers, Luria realized "that one would need a simpler system than the genes of a fruit fly if one wanted to verify experimentally the predictions of [his] theory." He describes his fateful introduction to bacteriophage through an episode quaintly called the "trolley-car accident."

> The tram I regularly took to work was almost as regularly stopped by electrical failures. One day, sitting in the paralyzed conveyance, I engaged in conversation with a fellow whom I knew by sight. He turned out to be a bacteriologist named Geo Rita, now professor of virology in Rome. My head that day was filled with Delbrück's articles so we talked of bacteria and genes and radiation, and when the

tram's electric current returned I went with my new friend to his lab to chat some more. He was at that time sampling the water of the Tiber for dysentery bacilli, using as a test the presence of something called bacteriophage, a virus-like parasite of bacteria whose name I had never even heard. Between bacteriophage and myself it was love at first sight. . . . Thus bacteriophage, Delbrück, and I were somehow brought together in that winter of 1938.

In 1938, Luria obtained a fellowship from the Italian government to study in the United States, prompting a departure from his native land that was hastened by Mussolini's formal alignment of fascist Italy with Nazi Germany. Following a brief sojourn in Paris, he arrived in New York in September 1940 where he took up his fellowship position at Columbia University. Within a few months, he met Delbrück and they immediately began their famous collaboration that resulted in the formation of the "phage group" and which culminated in their sharing the Nobel prize with Alfred Hershey years later. Luria spent the following summer at Cold Spring Harbor and of course he attended the 1941 symposium. Following a brief stint in Delbrück's laboratory, where "we explored, without much success, a number of possible approaches to phage multiplication," Luria was offered and accepted a faculty position at Indiana University in Bloomington in 1943. He recalled that "by some historical anomaly the University was then a leading center of genetics, prominently represented by Ralph Cleland, and Tracy Sonneborn, with H. J. Muller himself joining them in 1946."

Like Hollaender, Luria employed UV radiation extensively "as a microsurgical tool to analyze the properties and growth of bacteriophages." He served as Jim Watson's Ph.D. thesis advisor at the University of Indiana and as his mentor in succeeding years. In the late nineteen-forties, Luria discovered a phenomenon that he called *multiplicity reactivation*. He observed that UV radiation inactivated the ability of phage to yield progeny after infecting *E. coli*, a result that we now understand as consequential to many types of damage to the phage genome. He made the further observation that if the bacteria were infected with a very high titer of UV-irradiated phage, some viable phage progeny did indeed result. It is now known that when cells are infected with multiple phage genomes that have suffered DNA damage, recombination between the damaged genomes can reconstitute normal ones. Although not dependent on the direct biochemical repair of the DNA damage, multiplicity reactivation represents a form of cellular recovery from genomic injury by genetic recombination. Thus, its discovery might have provided another avenue for thinking in general, if not specific, terms about cellular events that contribute to the restoration of genomic stability after genomic insult. Here is what Luria had to say about DNA repair phenomena in general in *A Slot Machine, A Broken Test Tube*.

> Yet as early as 1946 I made a finding that was destined to open up a new insight on how the stability of DNA is achieved. I had long known that bacteriophage, when exposed to ultraviolet light, was killed in a straightforward way, the number of "kills" being directly related to the amount of light used. But Max Delbrück had noticed certain unexplained irregularities. I decided to look into this because I smelled some interesting possibilities. Sure enough, the cause of the trouble turned out to be something quite unexpected. What I discovered was that when two or more "dead" phage entered the same bacterial cell, they often became alive again

and produced normal live progeny. This was the first example of resuscitation or *reactivation of* cells or organisms that had been damaged by radiation. I interpreted the reactivation, correctly, as a result of genetic recombination. Two or more phages, if they were damaged in different genes, could by genetic exchanges like those recently discovered by Hershey and Rotman reconstitute an undamaged, completely normal phage.

And so I put Jim Watson, my first graduate student at Indiana, to repeat the experiments using X-rays instead of ultraviolet rays as the killer radiation. His results came out as I expected: with X-rays there was much less reactivation of dead phage than with ultraviolet light, because X-rays not only affect the DNA units chemically, like ultraviolet light, but can also break the chains, and broken chains are not expected to rejoin.

The discovery of reactivation of irradiated phage immediately started a flurry of activity in the study of repair of radiation damage. It turned out that the repair of damaged phage by mutual help that I had discovered was only one special case of DNA repair.

Judson puts a slightly different spin on Luria's interest in multiplicity reactivation. He recounts that the discovery of the phenomenon:

Suggested to Luria that the genetic substance of the phage consisted of different bits, and that inside the bacterium these separated, multiplied independently, and were pooled. Luria was on the trail of a number of discoveries that have since proved highly instructive; but the events by which phages grow within bacteria are more complicated than he then envisaged, so that within a few years his simple version of a gene pool had to be abandoned. Watson's X-ray project turned out to be a cul-de-sac down Luria's detour.

Watson's studies on multiplicity reactivation after exposure of phage to ionizing radiation culminated in a "safe" but "dull" Ph.D. thesis. Judson recounts that, true to character, Delbrück suggested to Watson that he was "lucky his thesis was boring; otherwise he might have to follow it up instead of having time to think and learn." One shudders to think of the implications of Jim Watson having directed his future career to radiobiology instead of molecular biology! In fact, his thesis work made such a limited impression on him that when I wrote to him to determine the title of his thesis he frankly replied that he did not recall. "My memories of my Ph.D. thesis research are now very limited," Watson wrote. "Several years ago, hoping to write some autobiographical pages about my Bloomington years, I began looking at more than 45-year-old notebooks. But they failed to reveal the exact ups and downs of my thoughts about the direct and indirect effects of X-rays as revealed by phage survival curves."

Parenthetically, in correspondence with Walter Harm in 1996, he pointed out to me that in 1956 Jean Weigle and Giuseppe Bertani at Caltech reported considerable levels of multiplicity reactivation of X-irradiated phage if the irradiation was applied *after* the phage DNA had entered the cell. Like Watson, they did not observe the reactivation if the phage were irradiated prior to infection.

Among the many experimental observations that emerged from Luria's laboratory at that time was one that came to hold special interest for DNA "repairologists." By dint of what was called the "phage treaty," Delbrück was able to persuade the entire phage community to focus on a particular set of phages called T (for Type) 1–7. According to Delbrück's biographers, Ernst Peter Fischer and Carol Lipson, this particular set of phages was chosen primarily because the plaques they formed on agar plates were easily visualized and could be readily quantitated.

> Once this decision was made, Max held to it firmly. He would not look at results of experiments done with strains other than those included in the treaty. Though Max never found this easy, he felt that a good researcher must tame his curiosity.

Luria noted that phage strain T4 was about twice as resistant to inactivation by UV radiation as the closely related strains T2 or T6. The systematic exploration of this observation by others culminated years later in the discovery of a DNA repair enzyme involved in a novel mode of excision repair. To this day, our understanding is that this enzyme is encoded uniquely by the genome of phage T4 and that of one other organism, a highly UV-radiation-resistant bacterium called *Micrococcus luteus*. We shall return to the history of the discovery of this phage T4-encoded enzyme in a later chapter.

Salvador Luria's contributions to the emergence of the DNA repair field have probably not been sufficiently appreciated, perhaps because that was never his primary research interest. His statement that the discovery of reactivation of irradiated phage "immediately started a flurry of activity in the study of repair of radiation damage" is an exaggeration. Regardless, the term "reactivation" that he and Delbrück apparently coined was an important conceptual and idiomatic prelude to the term "repair." As you will witness in the next chapter, Luria's studies on multiplicity reactivation were quickly followed by the independent discovery by his postdoctoral fellow Renato Dulbecco while in Luria's laboratory at the University of Indiana, of the first true DNA repair mode, which, based on Delbrück's suggestion, came by the name of *photoreactivation*. And as just indicated, his observations on phage T4, later dubbed *u*-gene reactivation, heralded the discovery of a particular mode of excision repair of UV-irradiated DNA.

In brief, the period from the mid nineteen-thirties to the late nineteen-forties may be characterized as one during which not only the field of DNA repair and mutagenesis emerged, but also much of what we now designate as the discipline of molecular biology. It was evident that agents that damaged the genetic material of cells elicited mutational changes. Additionally there were hints that living organisms responded to genetic insult in other ways, some of which were recognized as a temporary inactivation from which they could somehow recover. But a combination of intellectual biases, and to a lesser extent, political influences, constrained the emergence of gene/DNA repair as an area of investigative inquiry in parallel with other aspects of gene function. Genes were presumed to be made of proteins and to be intrinsically stable. There was no imperative to consider them at special risk to environmental or spontaneous damage, and hence in need of special biochemical perturbations. Mutations were considered to be rare events that were of enormous pragmatic value for genetic studies, but their mechanism of origin was not obviously experimentally tractable. Recovery after exposure to X rays and UV light was an anecdotal phenomenon at best, and at worst the province of government scientists who were primarily intent on gleaning useful biological applications for the militaristic use of radiation, a task for which they were lavishly supported. So, the first direct experimental evidence for DNA repair did not emerge until just before the middle of this century, and it was not until almost a decade later that the term was confidently and unambiguously incorporated into the lexicon of molecular and cellular biology.

A literary note to end this intentionally brief exploration of the earliest awareness of DNA repair was brought to my attention by my colleague and latter-day friend, Evelyn Witkin. Shortly after this chapter was completed Witkin shared with me this "delicious 17th century quote—which seemed to me to be a *really* early hint of DNA repair."

> If therefore Nature (through a penury or superfluity of materials, or other causes) hath been so unfortunate as at sometimes to miscarry: her dexterity and Artifice, in the composition of many, ought to procure her a pardon for such oversight as she has committed in a few. Besides there is often so much ingenuity in her very disorders, and they are disposed with such a kind of happy unhappiness, that if her more perfect works beget in us much of delight; the other may affect us with equal wonder.

CHAPTER 2

Let There Be Light: The Discovery of Enzymatic Photoreactivation

WERE IT POSSIBLE to custom design DNA repair mechanisms, perhaps the simplest and most efficient would be those in which specific chemical alterations in DNA were directly reversed by single-step enzyme reactions. Ideally, such reactions would be catalyzed by stable monomeric proteins with no requirement for exogenous cofactors. In fact, this biochemical paradigm for restoring the integrity of damaged DNA exists in nature. I refer to it as DNA repair by the direct reversal of damage. By curious coincidence, one of these relatively simple biochemical reversal processes, called enzymatic photoreactivation (or simply photoreactivation), was historically the first form of DNA repair to be discovered. In the previous chapter, I commented on the popularity that UV radiation enjoyed during the late nineteen-thirties and nineteen-forties as a means of generating mutations in cells, especially in microorganisms. UV light at a wavelength of approximately 254 nm (the predominant wavelength that emerges from an ordinary germicidal lamp) is efficiently absorbed by DNA. This type of radiation is also emitted from the sun, whence it is mainly absorbed by ozone in the stratosphere. But UV radiation of longer wavelengths penetrates the superficial cells of our bodies such as those in our skin and eyes, and some of this radiation is appreciably absorbed by DNA in these cells. So, aside from its convenience as an experimental tool, cellular responses to DNA damage caused by UV radiation are directly relevant to understanding the effects of a potent and prevalent natural mutagen and carcinogen, sunlight. Skin cancer is the commonest form of cancer observed in humans.

Among the many photochemical reactions that result from the absorption of UV radiation by DNA is one in which adjacent pyrimidines (C or T) in the same DNA strand undergo a chemical modification that results in their covalent joining or dimerization. Hence, these photoproducts in DNA are called *pyrimidine dimers*. Their discovery as products of the interaction of UV radiation with DNA represents an important historical landmark (that I shall recount in some detail later in the chapter) because it yielded the chemical nature of a specific substrate for the repair of UV-irradiated DNA, and this in turn greatly facilitated studies on several DNA repair mechanisms. It is probably no exaggeration to suggest that the multiple mechanisms by which pyrimidine dimers are removed from or tolerated in the genome of cells represent the most intensively investigated of all the known cellular responses to DNA damage. Since UV radiation derives from the sun, pyrimidine dimers are presumably as old as DNA itself. So, it is probably also accurate to suggest that this form of DNA damage evoked the evolutionary selection of multiple diverse repair mechanisms to contend with its potentially lethal and mutagenic effects. The morbidity and mortality that cells suffer in the presence of pyrimidine dimers derives from their interference with normal transcription and replication of DNA. Photoreactivation is one of several mechanisms by which pyrimidine dimers can be enzymatically removed from DNA. Specifically, during photoreactivation, dimers are split such that the individual component pyrimidines are restored to their normal chemistry and the DNA is returned to its native conformation and function.

The biological phenomenon of photoreactivation was discovered by Albert Kelner in 1948 and independently by Renato Dulbecco at about the same time. As you will see, some of the events surrounding these discoveries are as rich in melodrama as a Hollywood movie script. The historian's dream is to have (exclusive) access to firsthand written accounts of relevant events. My own aspirations in this regard came close to fulfillment when I was researching this era in the history of DNA repair. Adelyn Kelner, the late Albert Kelner's wife, informed me that "my husband was meticulous with his notes and record keeping." This turned out to be an understatement. Among other written legacies of his personal and professional life, Kelner left over 200 pocket diaries. He also filed away for posterity almost every draft of every major talk he delivered and every manuscript he wrote. In late 1961, Claud S. (Stan) Rupert (whose special contributions to the photoreactivation story will be recounted in detail presently) was preparing to write a comprehensive review article on photoreactivation and contacted Kelner for detailed information about his (Kelner's) discovery. For reasons that will become clear in the course of this chapter, Kelner was psychologically inclined to provide a long and detailed accounting of these events, events that he apparently had not shared with anyone else, except perhaps his wife and closest personal friends, during the preceding 13 years. In an elaborate epistle to Rupert (which took Rupert quite by surprise since this highly personal document was not quite the sort of response he anticipated), Kelner recounted in considerable detail the events that led him to the discovery of photoreactivation and much of his emotional state surrounding these events. I was fortunate enough to obtain this remarkable document from Stan Rupert.

Albert Kelner came from humble beginnings. Born in 1912 into a poor family, he was stricken with tuberculosis of the bone in his early teens. This affliction required frequent hospitalization and left him with a permanent and pronounced limp. His left shoulder was also affected and this seriously interfered with his considerable talent for the violin. Kelner's charming wife, Adelyn, with whom I had many poignant and interesting conversations, is convinced that had he not developed an early interest in biology, Kelner would have directed his career to music. "We might have starved as a family," she confided laughingly, "but that's besides the point—he was a very fine musician." Adelyn always spoke of her late husband with immense affection and it was evident that she was deeply devoted to him and that they enjoyed a close relationship. Indeed, she proposed marriage to him. "It was a whirlwind thing," she told me. "I was working in Washington, D.C. at the time and took a brief holiday in upstate New York. I met him in September during the Labor Day vacation. I proposed to him in October and we got married in December!"

Frequent hospitalization forced Kelner to forego much of his formal high school education, but persistent and diligent informal study ultimately gained him a full scholarship to the University of Pennsylvania based on his outstanding entrance examination scores. He acquired his bachelor, master, and doctoral degrees in a total of seven years.

Alexander Fleming's published discovery of penicillin in 1929 lay relatively dormant in the literature until it was resurrected by the noted English pathologist H. W. Florey (later Lord Florey and provost of Queen's College at Oxford University) in the late nineteen-thirties. During World War II, penicillin was mass-produced in the United States and an intense search for new antibiotics was launched. Between 1943 and 1946, Kelner was part of a research team at the University of Pennsylvania involved in such a research effort. His particular scientific focus was to develop ways of mass-screening various microorganisms for antibiotic production. He was also very keen on investigating other effects of the excretory products of microorganisms. He wrote to Rupert that:

> I thought excretory products might have specific growth-stimulating, growth-distorting or other effects, not only against bacteria but also against all sorts of living things. While developing such a method of screening I peddled my ideas around several laboratories hoping that they would help land me a job.

Kelner's peddling struck a receptive chord with Milislav Demerec, director of Cold Spring Harbor Laboratory during the period 1941–1960. A noted microbiologist, Demerec assembled the famous collection of T phages (mentioned in the previous chapter), which so effectively served early molecular genetics. Judson tells us that "[Jim] Watson remembered Demerec as the man who went around switching off lights and who refused to make inessential repairs like replacing a broken toilet seat." Demerec was intimately involved in the war effort to mass-produce penicillin in the United States and was also keen on exploring the notion that many microorganisms (including *E. coli*) might be mutated to antibiotic-producing forms. Supported by funds from a commercial company interested in antibiotic production (the biotechnology industry was apparently alive and well even in the pre-recombinant DNA era), Demerec extended an invitation to Kelner to join him at Cold Spring Harbor in order to screen mutants of *E. coli* for antibiotic production.

Kelner's discovery of photoreactivation is considered in many quarters to be serendipitous. The history of science is replete with examples of this quirk of human endeavor. Perhaps the most celebrated is the discovery of penicillin by Alexander Fleming, the very field in which Kelner began his career. But in fairness to the late Albert Kelner (who died in 1994), I do not share the view that his discovery was at all accidental. The word serendipity denotes engagement in a search for a particular objective and totally by chance obtaining an outcome that is very different, and typically not at all unpleasant. I believe that the word owes its origin to an old fairy tale entitled *The Three Princes of*

Serendip (Serendip is the former name of Ceylon). In this story, the principal characters, three princes in pursuit of a princess's hand in marriage, encountered examples of unexpected and unintentional good fortune. The version of this tale that I read chronicled that one of the three princes was searching for a lost arrow and instead encountered a beautiful maiden whom he wedded! Certainly a very different (and very pleasant) outcome from the task of searching for an arrow! Judge for yourself whether or not Kelner's discovery belongs in this category of pure and unadulterated chance.

Upon accepting Demerec's offer, Kelner moved to Cold Spring Harbor Laboratory—a place that, like most scientists blessed with such opportunity, he took to like a duck to water. A short time after beginning his mutational studies, for which he initially used UV radiation as a mutagen, Kelner noted and became increasingly obsessed with a quantitative postirradiation survival problem, which ultimately led to his now-celebrated discovery. He confided to Rupert that:

> Cold Spring Harbor was very stimulating and I fell in love with microbial genetics immediately, even though the original purpose of my going there to work side by side with a master was not to be achieved. My first task was to irradiate *E. coli* with UV light to induce mutants, and from the first experiment in October 1946 I ran into difficulty with the reproducibility of survival rates. Not that the curves were abnormal or not good straight lines on semilog plots, but that the level of accuracy required for my purposes was apparently greater than that required by previous workers with UV radiation. I would irradiate a suspension, assay an aliquot for survival, storing the remainder of the suspension at 5°C until the assay plates grew. Guided by knowledge of the exact titer of the particular irradiated suspension, I would inoculate at one time 200–400 plates so that each plate had from 20–50 colonies. Then the thousands of colonies were tested at one time for antibiotic activity. Fewer than 25 colonies/plate made the test inefficient, more than 50 made it inaccurate because of crowded colonies. Thus, I needed a suspension of irradiated cells whose titer was accurate to about ±25%. But irradiation with the same UV dose two days apart gave variations exceeding this limit. All Demerec could advise was more "care", use of a voltage stabilizer, different methods for stirring suspensions, etc. I spent weeks trying to perfect my technique—but to no avail. By October or November of 1946 I had acquired a healthy disrespect for the implications of quantitative exactness of the beautiful UV survival curves in the literature.

Stan Rupert suggested to me that Demerec's insinuations that Kelner was technically sloppy provoked a (well-known) stubborn streak in him which initially motivated him to persist with these experiments more out of dogged pride in getting them right than out of genuine intellectual curiosity. Regardless, at about this time Kelner elected to include fungi in his antibiotic studies because "the occasional *E. coli* antibiotic experiment which worked showed no clear cut antibiotic-producing mutants." He also decided to extend his mutagen treatments to include X rays. In experiments preparatory to antibiotic tests with *Actinomyces* mutants, he observed that bacterial suspensions irradiated with UV light died during storage at 5°C. So, in between the antibiotic experiments he embarked on a lengthy series of studies aimed at understanding the possible causes of death of UV-irradiated *E. coli* (and later the fungus *Streptomyces griseus*) stored in the cold. Once again he was plagued with the persistent observation that individual plate counts from UV survival curves were highly variable, differing at any single dose by 5–10-fold, even though "if you averaged the plates you obtained points all falling on a good straight line on the semilog plot of the UV survival curves." In a passing allusion to the antibiotic experiments (in which he was becoming increasingly less interested), Kelner told Rupert that:

> Actually it wasn't necessary to do these UV radiation experiments for I knew I could obtain mutants reliably with X rays, and though the X-ray machine used was in the Memorial Hospital 30 miles away in New York, I enjoyed the trip, and so far as the antibiotic work went would have saved time in the long run by abandoning UV light as a mutagen. But I continued with the UV, trying to make it behave, added nutrient broth to the suspensions after irradiation, and in other ways trying to make conditions of the experiment with UV comparable to the X-ray experiments.

On October 21, 1947, Kelner entered the following notation in his laboratory notebook under the heading "Apparent recovery of *Actinomyces* after ultraviolet irradiation":

> [*Streptomyces*] *griseus* UV'd, assayed immediately, then suspension put into icebox, and assayed at regular intervals to detect possible changes in [numbers of] apparently viable cells. Conclusion—very irregular count. There was a 10,000-fold difference between 4 and 24 [hours of postirradiation storage].

In his lengthy discourse to Rupert written 14 years after the event, Kelner wrote that:

> I don't remember just what I thought at that time. Besides the usual first thoughts of a mistake in arithmetic and dilutions, I probably thought I had an exaggerated type of postirradiation change such as mentioned in passing in some of Hollaender's papers. I was very busy during the time continuing with my antibiotic work and stuck in a recovery experiment every once in a while, when I could. About this time [December 1947] I was looking for new research problems, both because I had become a little dubious about the antibiotic research, and because I was hoping that maybe Demerec would keep me at Cold Spring Harbor longer. The grant [from the commercial antibiotic company] I understood wouldn't last much longer. I was anxious to engage in the sort of intellectually satisfying fundamental problems the people around me were doing. Also, at that time the world was suffering from its first wave of anxiety about the dangers of radiation to the human race and I hoped this research would lead to amelioration of this threat, besides being of great theoretical interest. The idea thrilled me, and I had moments of great elation.

A remarkable feature of this retrospective accounting to Rupert was Kelner's personal and at that time very private conviction that he was onto something important. He was confounded by the irreproducibility of his observations of significant, sometimes even massive, post-UV recovery of fungal spore viability, but he was beyond doubting that this reflected anything but some sort of meaningful biological phenomenon. He told Rupert:

> I wasn't really worried about the irregularity of the recovery observed. I knew recovery was real, but the "factor" causing recovery was still unrecognized. Once the factor was discovered, the irregularity would disappear. Neither was I concerned over the absence of recovery after X rays or apparently in bacteria. I thought *griseus* just happened (perhaps because of peculiar metabolism or temperature growth requirements) to be particularly suited to recovery under the particular conditions of my experiments, but that once we knew the mechanism in *griseus* we could devise means for recovery in all cells, even mammalian cells, and after X rays

as well as ultraviolet. Demerec never gave the idea more than mild clucks of lukewarm interest, and let me know that the antibiotic work wasn't proceeding fast enough, and I ought to be publishing papers on it, and so recovery continued as a strictly sideline research, one which however, occupied more and more of my thoughts.

It is relevant to consider some of the details of how Kelner did his UV irradiation experiments. He irradiated a suspension of cells, added nutrient broth, chilled the culture, and then streaked out precisely measured aliquots on plates. The streaking operation took about an hour, during which time the plates were of course laid out on the bench top, and hence were subject to any perturbations that might be operative in the immediate laboratory environment. Kelner noted in early 1948 that the plates that were the last to be streaked generally showed the greatest recovery. "I therefore suspected that exposure to room temperature influenced recovery and investigated the effect of holding seeded plates at room temperature for various intervals on apparent recovery," he told Rupert.

For much of the remainder of 1948, he continued to explore systematically the effect of postirradiation temperature on recovery and generated a series of curves in which survival of UV-irradiated cells was plotted as a function of the postirradiation temperature. He wrote to Rupert:

> While the room temperature recovery was consistent, the temperature-recovery curve looked a little more dubious. However, I was confident of being on the right track. And I talked to people about the data I was getting and the problem, especially to Luria who was staying at Cold Spring Harbor for the summer, together with his students Dulbecco and Watson. There was mild interest, and little comment.

I will return to the significance of these discussions presently, because as mentioned earlier, photoreactivation of UV-irradiated phage T2 was independently discovered by Renato Dulbecco in Salvador Luria's laboratory at the University of Indiana in Bloomington at about the same time that Kelner ultimately solved his recovery phenomenon. By the summer of 1948, Kelner was rapidly exhausting Demerec's patience for the recovery experiments, and more rapidly running out of time before the grant that supported the antibiotic experiments was to terminate. He decided to make "a valiant effort" to determine the nature of the temperature-dependent recovery by accurately controlling temperature over a narrow range. In order to do this experiment, he moved to a laboratory (called the Jones laboratory) in a different building at Cold Spring Harbor. He told Rupert that:

> Jones was in a building nearer the beach and it was pleasant to watch the summer visitors—it was brighter and cheerier. It also had an adjustable "cool room," at one time used for summer work with *Drosophila*. I gathered all the water baths I could from everyone who had any to spare, and put some into the "cool" room in Jones. This was a windowless room which could be adjusted to a minimum temperature of 14°C. Water baths inside this room I adjusted to 19°C and 25°C, which was to be my accurately controlled room temperature. For temperatures over 25°C, I used water baths in the open lab (Jones). Now I had accurately controlled temperatures at 0, 5, 14, 19, 25, 30, 35, and higher.

Over the course of the next several weeks, Kelner's laboratory notebook painfully documented the disaster that struck his temperature-dependent recovery hypothesis, for the results of his new experiments made absolutely no sense to him. He recalled to Rupert that:

> I remember feeling very sad, confused and bewildered. The beautiful temperature-recovery curve I thought I had had fallen apart. There seemed so little time. I was spending most of my time writing a big paper on all the antibiotic work I had done during my years at Cold Spring Harbor, and hurrying to finish it, hoping it would get me a job. Demerec had given me absolutely no encouragement that I could stay beyond the fall—it would depend on whether he could get another grant, and he didn't think he could. There were wind-up experiments to do for the antibiotic work. Moreover, Demerec seemed unhappy and angry at me for moving to Jones.

However, by early September of 1948, light (in the metaphorical and literal sense of the word) appeared at the end of the tunnel. Here are some illustrative notebook and diary entries at that time.

> <u>Sept. 2, 1948</u>. Must try effect of light-sunlight-diffused light on recovery (infra red). Our 35 [°C] in direct light.
>
> <u>Sept. 4, 1948</u>. Noted that the 35° water bath was in full light on the lab table. Noted too that in old exp. at room temperature, in which recovery was greatest, the spores were in transparent bottles on the lab shelf exposed to day light.

Kelner's shift in focus to light exposure instead of temperature as an experimental variable was a fundamental turning point and there is little to suggest that this was purely by accident. During the course of the next few weeks, he systematically examined the effect of light on the recovery phenomenon and obtained unequivocal and reproducible evidence that when UV-irradiated spores of the fungus *Streptomyces griseus* were subsequently exposed to visible light, they sustained an enormous level of recovery compared to controls that were shielded from light exposure. Temperature had nothing to do with the recovery. All of the experimental inconsistencies that he had doggedly and methodically pursued for the past several years could be explained by random exposure of some of his experiments to light. In particular, he was now able to fully reconcile his recent chaotic temperature curve with the fact that some of the water baths were used in the sun-exposed Jones laboratory, whereas others were used in the shielded "cool" room. He wrote to Stan Rupert:

> After the final conclusive experiment of Sept. 13 [1948], I could delay closing my laboratory no longer. I told Demerec right away about the visible light, and although he perked his ears up a trifle and some expression of interest crossed his face, there was little comment. Taking stock of the future, to see what I could do next, and realizing that I had no job prospect, I asked Demerec what was to be done. He said that I could stay through the winter and work with Bryson on a grant obtained for Bryson on antibiotic resistance. This I refused, for how could I bear to abandon the recovery work now? I made a bargain with him—let me concentrate on the recovery problem and I would guarantee to leave by May. There was enough money left from the old grant to pay my salary, and I would work without an assistant or purchase of new equipment, etc. Meanwhile I could continue looking for a job. Demerec said later that I could have continued with photoreactivation—but this is simply not true. The recovery work wasn't mentioned at all, and I knew I would have to confine myself to hurried sideline experiments while my main responsibility would be the antibiotics. Others have told me that I should have accepted [Demerec's offer] and worked on recovery anyway, deceiving Demerec as to what I was doing. I was too naive for this.

Kelner's wife Adelyn shared a letter with me that was contemporaneous with this exciting period. The letter was written to his former mentor at the University of Pennsylvania, Wesley G. Hutchinson. It dealt mainly with his antibiotic work, but Kelner concluded this communication by relating his obvious elation about his new discovery, muted by his increasing anxiety about future job prospects. Adelyn stressed repeatedly that Al was extremely worried about finding a job and earning enough to support his wife and new baby. We shall see presently that this anxiety significantly influenced his demeanor in resolving his impending confrontation with Salvador Luria. He told Hutchinson that:

> Things are going well with me, especially as far as research goes, though the future is still uncertain as ever. My time here is definitely up in May [1949]. I've decided no more antibiotic research for me unless it is part of a wide, general problem (and I may have to). Drove a hard bargain with Demerec to let me work on what I want to for my remaining months here, and though he is perhaps a little nettled, I'm sure that in the end he'll be glad, for I have a juicy, most exciting problem, which is yielding results at every experiment. So that despite the fact that I can give only part time to it I think that it will be more helpful for my career than my last six years of work.

Shortly before this correspondence, Kelner wrote to Salvador Luria, then at the University of Indiana in Bloomington, with whom, as already mentioned, he apparently had discussions during the previous summer about his perplexing recovery experiments when Luria was temporarily resident at Cold Spring Harbor. This letter is pivotal to the events that followed, because regardless of how much Luria knew or cared about the complicated results related to him by Kelner that summer, this correspondence, provided verbatim below, unfolded in graphic and unequivocal detail the explicit information that recovery of the viability of UV-irradiated fungal spores was dependent on subsequent exposure to visible light.

> October 30, 1948
>
> Dear Dr. Luria,
>
> As a veteran father, I can understand the very full life you and Zella [Luria's wife] must be leading since the arrival of the baby. I hope you and your family are thriving.
> There has been a rather exciting development in the research on recovery after irradiation, about which I talked to you last summer. I thought you might be interested in hearing about it, and I would appreciate your comments. Last summer I had been investigating the temperature-recovery relationship, and found that there was several hundred-fold to a thousand-fold recovery when the *Actinomycete* spores were stored in saline at about 15°C, and at 45°C, with no or little recovery at 0° and 25–37°C, the latter being about the optimum temperature range for growth of the organisms. The recovery-temperature curve thus had two peaks.
> I have discovered however, another factor which entirely overshadows in importance the temperature effect. This is irradiation of the ultraviolet-treated cells with nothing more than visible light. Under suitable conditions such irradiation will cause over 200,000-fold recovery, such a tremendous recovery that I feel that I have hit upon the key factor within the cell which can bring about recovery after ultraviolet (or X-ray?) treatment. This factor was investigated because I had noticed that suspensions stored on the laboratory shelf in the presence of diffused light

from the window had a far greater recovery than suspensions stored in a water bath (at approximately room temperature) which was partially shielded from light. Also because when I moved over to Jones toward the end of the summer, recovery of suspensions in the 35 water bath became high and variable. This turned out to be because the water bath had been placed in front of a window.

My plans are to (1) standardize conditions under which maximum and most rapid recovery will occur, so that I have something to work with, (2) determine whether the light is affecting something within the cells, or something in the menstruum (perhaps peroxides?), (3) investigate the effect of various wavelengths, in order perhaps to get a spectrum of the relative efficacy of various wavelengths on recovery. This may give me a clue as to what compound in the cell is being affected, (4) determine the generality of the phenomenon by studying recovery under standardized conditions of several actinomycetes, fungi, bacteria, and phage.

What do you think of all this?

Conditions for me are in as chaotic a state as ever. Demerec agreed to allow me to stay for this winter so that I could find some sort of job, but he wanted me to work on some problem he had gotten a grant for Vernon [Bryson] for (quite complicated isn't it?) work on resistance of acid-fasts to streptomycin. Had I agreed to work on this problem I could have stayed another year, but I bargained with him to the effect that if he let me work on the recovery problem, I would guarantee not to stay longer than May—until May because there was just enough money left from the Schenley grant to keep me until then. He agreed with a lot of scowls and frowns, and so I have this winter to look for a job, and work on this problem. There is no assistant however, or money for equipment, and so I have become an expert contriver of apparatus made of rubber bands, cardboard and Scotch tape!

The papers on the *Actinomyces* mutants, and antibiotic work are all in press; the main one will appear in the January issue of the J. Bacteriology.

I have become quite disheartened about the prospect for a good job, and am about ready to throw up the sponge, and go into some commercial laboratory. I did have some correspondence with Topping of the National Institutes of Health about the possibilities there. But nothing seems to be coming of it. Do you know anyone at the National Institutes of Health who might be interested in the sort of research I have been doing?

I had been hoping that if nothing more I could go down there as a guest investigator for a few weeks this fall, to do the spectrum phase of the research, a phase for which the equipment here is hopelessly inadequate. That would give me a chance to breathe the atmosphere of another laboratory for a change as well as a chance to get an important phase of the work done. If I don't succeed in accomplishing this at NIH, I'll try one of the laboratories in New York. Incidentally, in some of the applications I've been making I took the liberty of giving your name as a character reference. Hope this is all right.

Your comments will be much appreciated.

Sincerely yours

Albert Kelner

There is every indication that Kelner liked and respected Salva Luria well. He apparently viewed Luria very much as a mentor and he openly solicited his advice on the recovery findings (the full details of which are clearly revealed) and he relied heavily on Luria's good graces and considerable reputation in helping him secure employment. Imagine Kelner's shock and surprise when he received the following reply from Luria almost a month later.

> November 26, 1948
>
> Dear Kelner,
>
> You will be interested in knowing that Dulbecco has discovered, quite by accident, a phenomenon which may be the counterpart on phage of your discovery on bacteria—ultraviolet-inactivated phage is reactivated by visible light at a terrific rate—the conditions are extremely peculiar, and it will take several weeks to know where the radiation acts—Dulbecco has isolated pretty well the active from the inactive bands of light. For the time being it is not clear whether the action is on phage itself, on medium, or on bacteria. We shall keep you informed of any progress, and at the same time I'd like you to let me know if you have some result or idea that may help us. In about 2–3 weeks we ought to have enough data to give you a quantitative summary.
>
> I have made some inquiries concerning positions suitable for you. Would you mind sending me a brief biography (8–10 lines) and list of publications. It may help.
>
> Best regards, also to Mrs. Kelner.
>
> Yours
>
> Luria
>
> P.S. Danny is doing fine, passed the 12 lbs mark, doubling birth weight in 11 weeks.

Kelner immediately shared this letter with Demerec and others at Cold Spring Harbor. In his correspondence to Rupert 13 years later, he commented that:

> Demerec and the staff at Cold Spring Harbor, especially Barbara McClintock and Caspari were far more indignant and skeptical than I, and told me so. All the letters I wrote (after the first) to Luria were approved by Demerec. Actually I have the highest respect for Luria and Dulbecco and was glad to follow Tom Anderson's advice that the best thing to do about such a matter is to forget it. But after all these years it is proper to let someone else besides myself know what went on. For I believe that it has plagued my career ever since. Of course photoreactivation would have been discovered eventually (Professor Magni, Institute di Genetica, University of Pavia told me last year [in 1960] he was observing it in yeast when my paper was published) by somebody. And even maybe Dulbecco would have. But he certainly knew about my work before making his observation. You can imagine how I must have felt at the time, with no job, or opportunity to work, and anxiety about the future.

But Kelner did not "forget it." Urged by Demerec he began writing up his results for publication in the proceedings of the National Academy of Sciences, of which Demerec was a member, thereby enjoying the privilege of rapid communication to the journal. Perhaps primarily because he was busy with this writing, perhaps for other reasons, the specifics of which we shall never know, Kelner did not immediately respond to Luria's letter of November 26, a nuance of some significance, as we shall presently see. But his concern and unhappiness about this situation were heightened by a second letter from Luria written just prior to Christmas of 1948.

> December 23, 1948
>
> Dear Kelner,
>
> Because of the extreme interest that the photoreactivation [it would appear that Luria had already named this phenomenon] of phage will have for virologists, we have thought that Dulbecco should send a note to *Nature* briefly relating the facts. I

thought that unless you have already published your observations on bacterial resuscitation, you might like to send in a similar note. I am enclosing a copy of Dulbecco's note.

Dulbecco ran into photoreactivation in a most queer way, by forgetting to put off the fluorescent light on a table on which he had left a pile of plates with irradiated phage to incubate them at room temperature. Next day the top plate had 100x more plaques than the bottom one, and the intermediate ones had gradually different numbers. He has investigated the phenomenon very thoroughly from a physical point of view, isolating the effective wavelengths, etc. It is a most exciting thing, and I imagine that the bacterial phenomenon you discovered must also be such.

Please let me know how your plans are developing. There are chances that something suitable for your needs and interest comes to my attention soon, in which case I shall let you know.

With best regards and wishes for the holidays, I am

Yours,

S. E. Luria.

Unfortunately, Al Kelner died during the summer of 1994 and I was not fully able to get a firsthand measure of this apparently gentle and rather private man, nor to establish precisely what he thought and how he felt during the period between late December 1948 and January 15, 1949, when he replied to Luria's two epistles. When he finally did so he composed a masterpiece of professional sobriety and decorum in which he deliberately adopted a calm, reasoned, and forthright appeal to what was obviously a delicate and (at least for him) emotionally charged situation. As mentioned earlier, several people who recall that time, most notably his wife Adelyn, told me that securing Luria's good graces to help find a job was not a trivial motive in his demeanor. Yet he was clearly unwilling to capitulate on the important principle at stake for him—recognition and priority for his years of individual labor. As you shall see, gratifyingly for all concerned, this appeal struck a responsive chord in a presumably more than slightly embarrassed and somewhat chastened Luria.

January 15, 1949

Dear Luria,

I want to thank you and Dulbecco for sending a copy of the ms. It was indeed very gratifying to learn that light-induced recovery occurs also in phage, as I had suspected. (You will remember that in my letter of October 30 I mentioned that I planned to try my recovery experiments with phage, but of course that won't be necessary now.) Phage photoreactivation also makes more certain than ever that my feeling that the phenomenon is a general one is correct. There is nothing I should like better than to exchange information with you and Dulbecco; I intend to do so, and hope that we will both progress the faster for it.

However I want to first explain to you as frankly as I can some of my more personal reactions to your letters. And before beginning I know you will agree, that if our positions were reversed you would most certainly feel exactly the same as I do now. It is this: it seemed a most unusual, and almost impossible-to-believe coincidence that Dulbecco's discovery should have entirely independently been made precisely 3–4 weeks after I had written you the essentials of my findings. I do not imply the first impetus to Dulbecco's discovery (the pile of queer plates) was not wholly unplanned; but that my data certainly must have helped in the interpretation, in the exclusion of other possibilities, etc., etc. I remember from last summer how closely you two work together. Now light-induced recovery is certainly not an obvious phenomenon, for if it were then Hollaender, Latarjet, you or Dulbecco would have discovered it long ago. I'm sure plates have been exposed to light before. Nor does the phenomenon proceed obviously from the Hollaender-Kaufmann infrared studies; those dealt with mutations and you know yourself the other fundamental differences between their work and ours.

I cannot help feeling—and again I say that if our positions were reversed I am positive you would feel the same way—that my findings had influenced the discovery of phage photoreactivation, and I would have felt much better if my original discovery and its relation to Dulbecco's were mentioned in your ms. to Nature and in your discussions with others (such as Anderson, etc.).

What I am confident of is that in the excitement of Dulbecco's discovery, the influence of my findings may have been entirely unconscious and indirect.

I am sure this matter of which I have spoken so frankly will iron itself out, and we can discuss matters in a most friendly manner. Incidentally, Demerec has been exceedingly enthusiastic, helpful and sympathetic to me in this entire matter—both in its scientific and non-scientific or personal aspects.

My best regards to Zella and the new baby, and a happy and scientifically progressive New Year!

Yours,

Albert Kelner.

One cannot fail to be impressed by the apparent efficiency of the U.S. Postal Service between Cold Spring Harbor, New York, and Bloomington, Indiana in those days prior to the technological wonders of facsimile machines and electronic mail. Just two days later, on January 17, 1949, Luria received the letter from Kelner quoted above and immediately drafted a detailed response which, in contrast to his earlier correspondence, he formally copied to his graduate student, Renato Dulbecco.

January, 17, 1949

Dear Kelner,

I received this morning your letter of January 15. At first I was surprised at your reaction, but I must say that on second reading I saw your point of view and agreed with it. Dulbecco's observations came out in such an astonishingly independent way that the possible subconscious connection between your results and his observation never quite materialized in our minds as one of cause to effect. It must be recognized, however, that an influence of your original communication in formalizing the interpretation may well have occurred. I want to give you the full details of what happened, and then suggest a solution that may be satisfactory to you and Dulbecco both.

1. For several months, we had been puzzled by a lack of reproducibility of plaque counts in pairs of plates used in assaying the titer of irradiated phages. Tests of several kinds failed to give any explanation, and the observation was shelved as a nuisance. This was in September [1948]. The reception of your letter failed to suggest to me the obvious interpretation, that one of the two plates sometimes remained on top of the other on the table for an hour or more, and therefore received more light. Incidentally, this difference between assay plates only came up either in Cold Spring Harbor, with lots of diffused light, or after we installed here fluorescent lights directly on the lab tables. As a matter of fact, regular incandescent bulbs give out very little of the photoreactivating wavelengths.

2. While I was in New Haven November 10–18 Dulbecco was doing experiments on the effect of temperature on reactivation by multiple infection. In a series at room temperature (26°), at 33° and at 37° it came out that there was an excess of reactivation at 26°. In a second experiment (20° and 37°) there was an excess at room temperature again. He did a third experiment, comparing 26° incubator room and regular room, and in the latter one there was an excess. In thinking of possible differences he noticed that the plate that was on top of the pile at room temperature had the most plaques, and the lower ones had decreasing numbers. The pile of plates had been under the fluorescent light for several hours. At this point he remembered the difference between plates in pairs and tested for it. By the time he met me in Chicago (November 19) at a joint seminar with Szilard he had explained the difference.

3. Your letter [of October 30, 1948] arrived around November 1. I told Dulbecco about it, but he did not read the letter. We did at no time plan to test photoreactivation on phage. *A posteriori* and incidentally, the simplest test for phage reactivation would have failed, since phage is only reactivated in [the] presence of bacteria. I am perfectly sure that Dulbecco had no conscious recollection of your results, since I remember that I reminded him of them in Chicago. I think, however, [it] very possible that the process of interpretation was accelerated from having heard of your results a few weeks earlier. That he did not think of them consciously can easily be seen from his protocol of daily experiments, in which you can see that he was groping completely in the dark.

Luria must have been in a solemn frame of mind to have passed up the opportunity to recognize (and comment on) the pun in his choice of the word "dark"! In a more serious vein (which presumably reflects Luria's own demeanor during this difficult time), at this point in his letter Luria offered a remarkable gesture of capitulation in restoring priority for the discovery of photoreactivation to his younger colleague, and simultaneously extending a candid apology for his previous failure to acknowledge the obvious importance that Kelner attributed to his experimental findings. The letter continues:

4. In view of the above, I think it is only fair that you should have the complete credit for the first discovery of photoreactivation. My suggestions, which I want to submit to you for approval before anything is done (besides stopping publication of the note in *Nature*, which I have already done telegraphically), are the following:

a) Dulbecco's note could have the following paragraph inserted after the first one:

"The occurrence of photoreactivation of ultraviolet irradiated phage was noticed accidentally a few weeks after receiving a personal communication from Dr. A. Kelner that he had discovered recovery of ultraviolet treated spores of actinomyces upon exposure to visible light. My observation indicated the correctness of Dr. Kelner's suggestion that the phenomenon discovered by him may be of general occurrence for a number of biological objects."

Also, in the first paragraph, the word "discovered" on line 4 could be replaced by "observed."

b) If you consider this satisfactory, the note on phage could be sent on to publication, if you do not expect to publish your discovery soon. It is important to us to make the distinction of photoreactivation from reactivation by multiple infection known soon, since it may affect the mechanism of the latter reactivation, on which there are several papers in press. If, however, you plan to publish soon, Dulbecco agrees to delay publication of his observation until that time. Incidentally, we would appreciate your giving him permission to do so as soon as possible; most of his data, as you will realize have more relevance for phagology than for the mechanisms of photoreactivation, and that is what we are mainly interested in.

After all this on a technical level, let me personally assure you that we never had the slightest intention to capture priority from you, as our prompt willingness to abide by your decision proves it. You can imagine that we were very much upset by the possible consequences of photoreactivation of phage for the whole problem of the genetic interpretation of reactivation by multiple infection (and we still are in part). When Szilard and then Delbrück suggested that the thing should be announced quickly to keep other people from misinterpreting results, we did so, and at that time I wrote to you for your opinion. After failing to hear from you, I sent the note to *Nature*, without giving enough thought to the possible influence that your discovery may have had on the course of Dulbecco's work. As I already stated before, there was no conscious influence, and the possible subconscious one I failed to appreciate sufficiently.

I hope that my suggestions meet with your approval. Please do not let this apparent misunderstanding alter your good feelings toward us. If you had written me immediately there would have been no such complication.

With best regards, also from Dulbecco. I remain,

Yours sincerely,

S. E. Luria.

cc R. Dulbecco

Luria's comment about failing to obtain a response from Kelner to his letter of November 26, 1948 is significant. As he pointed out, having specifically informed Kelner about the observations in his own laboratory, he (Luria) was focused on the imperative of forging ahead with a communication to *Nature*, an imperative that was apparently reinforced by his discussions with Max Delbrück and Leo Szilard. The rationale for this haste is amply documented in the first paragraph of the paper that was eventually published by Dulbecco in *Nature*: "Since [the] phenomenon [of photoreactivation] may cause serious misinterpretation of results obtained in working with irradiated phage, it may be useful to report it at this early stage of its investigation."

It is evident from Luria's January 17, 1949 letter to Kelner, as well as from the perspective adopted by Dulbecco in his *Nature* paper, that Luria was not especially interested in the recovery by phage or bacteria from the inactivating effects of UV radiation, even though in his earlier letter written just prior to Christmas of 1948 he described Dulbecco's result as "a most exciting thing." As recounted in the previous chapter, at that stage of his career, Luria's focal point of research was the phenomenon of multiplicity reactivation of UV-irradiated phage, a phenomenon that he hoped might provide insights into the mechanism of phage replication in bacteria. Indeed, neither the episode of his extensive correspondence with Kelner, nor Dulbecco's independent discovery of photoreactivation were recounted in his autobiography *A Slot Machine, A Broken Test Tube*. And Dulbecco's *Nature* paper was published without Luria as a coauthor.

To close this particular chapter of the photoreactivation story, here is Kelner's response to Luria's gracious letter of conciliation and apology.

January, 20, 1949

Dear Luria,

The solution you suggest is a most fair and decent one, and if the insertion and emendation you suggest are included in the note to *Nature* I of course give my whole-hearted approval for the immediate publication of Dulbecco's findings.

At Demerec's suggestion I had submitted a manuscript for publication some weeks ago, and perhaps if possible you might also want to mention this paper as "in press, Proc. Nat. Ac. Science." Although this is not too important a point, and it would not be worth delaying publication of Dulbecco's ms. to include this reference.

This is a hasty letter for I wanted to write immediately to go ahead with publication of the note to *Nature*. I'll send a longer letter soon, with a copy of my manuscript. I'm very glad to have this affair off my mind and look forward to discussing the scientific points of this phenomenon.

I agree that photoreactivation is an important discovery for phagology. Indeed one reason I have not discussed the phenomenon with very many people is that I wanted to give your laboratory a chance to work out the problem and announce your findings as they relate to reactivation in general yourselves.

With best wishes, and thanking you and Dulbecco for your honest and sincere reaction to my letters.

I am

Sincerely yours,

Albert Kelner.

Kelner's priority for the discovery of photoreactivation was formalized when his paper in the proceedings of the National Academy of Sciences (officially communicated by Demerec on December 5, 1948) was published on February 15, 1949, several months before the appearance of Dulbecco's letter to *Nature* on June 18 of that year. The latter publication included all the amendments that Luria promised and did indeed cite Kelner's unpublished (at that time) paper. There is little doubt that this discovery resurrected Kelner's career, about which he was so concerned. Soon after completing his sojourn at Cold Spring Harbor Laboratory, Kelner obtained a Special United States Public Health Service Fellowship to work at the Biological Laboratories at Harvard. During this period, he published an article on photoreactivation in *Scientific American*, a significant accolade accorded to scientists by invitation in order to draw public attention to important biological phenomena, and soon thereafter he was recruited to the faculty at nearby Brandeis University in Waltham, Massachusetts, which had just opened its doors, and where he spent the remainder of his scientific career.

Following the publication of his initial experiments, Kelner thoroughly researched the literature for documented evidence that photoreactivation might have been previously observed. Interestingly, he discovered that as early as 1904 the German botanist Ernst Hertel had observed that UV light inhibited the movement of protoplasm in *Elodea* leaf cells more strongly in the dark than in the light. He also uncovered the fact that in 1933 two German physicists, Rolf Hausser and Reinhold Oehme, reported that visible light inhibited the browning action of shortwave UV radiation on banana skin, and that in 1941 Douglas Whitaker at Stanford University had reported that white light counteracted the growth-inhibiting action of UV light on the alga *Fucus*.

What is the retrospective view from the Luria camp about the events of the summer of 1948 and the period immediately thereafter? Lamentably, as was the case with Al Kelner, Luria was no longer alive when I began this historical inquiry. But Renato Dulbecco and I enjoyed an interesting attempt at reconstructing some of the events of almost half a century ago. We met in Francis Crick's opulent presidential office at the Salk Institute (which Dulbecco borrowed for the occasion since his own office was temporarily usurped by a sabbatical visitor at the Salk), which majestically overlooks the Pacific Ocean for as far as the eye can see. Like Luria, Dulbecco attended medical school at the University of Turin, but he did not know him well at that time since Luria was a class ahead of him. However, they were both close friends of a fellow medical student, Rita Levi-Montalcini, who incidentally, later became a highly celebrated neurobiologist in the United States. In his autobiography, Luria referred to his good friend Rita as "the queen, because of her impeccable dresses and regal manner." Remarkably, all three friends and fellow medical students would achieve the singular distinction of winning Nobel Prizes—Dulbecco in 1975 together with David Baltimore and Howard Temin for their contributions to the biology of tumor-causing viruses, and Levi-Montalcini in 1982 together with Stanley Cohen of Vanderbilt University for their seminal work on nerve growth factor. As already mentioned, Luria attained this distinction in 1969, together with Max Delbrück and Al Hershey.

Aware of Dulbecco's burgeoning interest in genes and genetics, Levi-Montalcini suggested that he have a chat with Luria, who happened to be visiting his native Turin in the summer of 1946. These discussions paved the way for the conversion of yet another Italian physician to the world of molecular biology. Luria suggested that Dulbecco (who in addition to his M.D. had almost completed the formal equivalent of a Ph.D. degree in physics) join his laboratory in Bloomington. Armed with a postdoctoral fellowship secured by Luria, Dulbecco arrived at the University of Indiana in 1947, just a year before Jim Watson joined Luria's laboratory as a graduate student. In addition to his outstanding contributions to cancer biology, a more recent but possibly less celebrated contribution that Renato Dulbecco made to science and to society was his public suggestion that the human genome be sequenced. Many regard this rallying cry as the initial fuel that ignited the extensive polemic that culminated in the formal establishment of the Human Genome Project. Here is the perspective that Dulbecco offered in an editorial in *Science* published in 1986:

> If we wish to learn more about cancer, we must now concentrate on the human genome. We are back to where cancer research started, but the situation is drastically different because we have new knowledge and crucial tools, such as DNA cloning. We have two options; either to try to discover the genes important in malignancy by a piecemeal approach, or to sequence the whole genome of a selected animal species . . . I think that it will be far more useful to begin by sequencing the cellular genome.

Dulbecco's recollections of his encounter with photoreactivation early in his scientific career were understandably vague. Aside from the fact that these events transpired a long time ago, he did not view them as especially significant in his career development, which very soon thereafter switched to animal viruses. Hence, the drama recounted in the correspondence between

Kelner and Luria was not at all vivid in his memory. When I suggested that, in retrospect, photoreactivation was the life blood of Kelner's scientific existence, whereas for Luria it was more an interesting coincidental sort of thing, he emphatically agreed. "Exactly," he told me. "And for me too, because in fact I didn't really pursue it at all." He had absolutely no recollection of any correspondence between Kelner and Luria and expressed no special interest in reading it 50 years later. Indeed, Dulbecco frankly admitted that while he certainly did not disdain history, he had no particular interest in the topic, and despite his close personal relationship with Luria and his wife Zella during his early years in the United States, he had not even bothered to read Luria's autobiography. "History doesn't interest me," he confessed laughingly. "It's important of course—but that's just the way I am."

When he joined Luria's laboratory he immediately set to work exploring aspects of multiplicity reactivation. "Luria wanted me to explore lots of different multiplicities and other conditions," he told me. He recalled that multiplicity reactivation was definitely the central experimental theme in the Luria laboratory at that time. He also agreed with the historical notion that Luria viewed multiplicity reactivation as an important clue to the mechanism of phage replication, and that he used UV radiation primarily as an experimental tool and had little, if any, abiding interest in recovery from radiation damage as a biological phenomenon. Indeed, when he first informed Luria of his discovery of photoreactivation, which he confirmed was made at a time when Luria was out of town at a meeting (in early November 1948 according to Luria's letter), Luria's immediate reaction was one of distress and concern that the light effect might have confounded the multiplicity reactivation data. As soon as Luria learned about Dulbecco's results he made him repeat several experiments in the absence of photoreactivating light to be sure that the multiplicity reactivation data had not been influenced in some way by light exposure.

As for the details of his discovery of the phenomenon, Dulbecco described the events to me as follows:

> I was doing multiplicity experiments of one sort or another. You know how it is, you have two plates for duplicate points and you put one on top of the other on the bench, wait until the agar solidifies, and then put them in the incubator. I noticed that the two plates never had the same number of plaques when we used UV-inactivated phage. I tried to think why that might be happening. One idea I had was that since there are two plates maybe one cools more rapidly than the other, so I tried to see whether incubating the two plates at different temperatures had any effect and it did not. Then one day I looked up and saw this huge fluorescent light just above the bench and I remembered having vaguely heard something about the effect of light from Kelner's work at Cold Spring Harbor. I was there in the summer of 1948. I never actually met him, but I knew that he was working with bacteria and that he had observed something interesting about the effect of light on UV-irradiated cells. That was significant, because once I recalled something about a light effect, looking up at this big fluorescent light I immediately thought about it—otherwise maybe I would not have. And so I did the obvious experiments and they came out very clearly. Jim Watson was around at the time and he was amazed at the really fantastic effect.

As Dulbecco recalled the subsequent events, Watson and Luria suggested that he write up the light effect on the reactivation of bacteriophage, which they decided to call *photo*reactivation. "I remember that there was a complication," he told me reflectively. "I wrote the paper and in the version that I put together I did not mention anything about Kelner's work because as far as I was concerned this was just hearsay. But after we sent it to *Nature* Luria told me there had been some sort of complaint about this and that we had to be sure to acknowledge Kelner specifically. So we did, and the manuscript was accepted." Dulbecco recalled nothing about the nature or source of such a complaint. He remembered presenting the results of his early studies and a few other experiments that he did on photoreactivation of phage at a meeting in Oak Ridge in 1949 (see below), but had no recollection of meeting Kelner there.

Dulbecco's comments led me to the conclusion that there was never any question in his mind about the issue of priority for the discovery of photoreactivation. He stumbled onto this DNA repair phenomenon quite accidentally during the course of his multiplicity reactivation experiments, and his vague recollection of what he had heard about the effect of light at Cold Spring Harbor the previous summer quickly focused his attention on this experimental variable. As a recent, young postdoctoral visitor to the United States still struggling with the English language and intent on finding his scientific and personal feet so to speak, he was apparently quite removed from the political implications of the timing of his experiments with those of Kelner. Both he and Luria found the observation of photoreactivation interesting, but Luria was chiefly concerned about its potential for interfering with their multiplicity reactivation work and hence deferring priority to Kelner was in the final analysis not especially difficult. Nor did Dulbecco consider this gesture unusual for Luria, whom he holds in very high regard and considers to be have been extremely fair and open in general. The fact that Dulbecco was the sole author of the *Nature* letter on photoreactivation of phage was, according to Dulbecco, typical of Luria's generosity in according full credit to his students and fellows concerning work with which he had no direct involvement. "He had nothing to do with that work," Dulbecco remarked. "He wasn't even there when it happened."

As our lengthy discussion wound down, Dulbecco and I agreed that the spirit of cordiality and sensitivity symbolized by the "Luria-Kelner affair " was perhaps more the rule than the exception a half century ago, and that science in general was conducted in a more open and less paranoid atmosphere during that time. Dulbecco recalled that "at Caltech Max Delbrück's fundamental tenet was that every experimental result had to be made known to anyone and everyone who wanted to know—anywhere in the world," and he acknowledged that he had been privileged to have grown up in a wonderfully innocent period in the history of biology.

With the wisdom of hindsight, there are providential elements of Kelner's experiments that clearly helped his discovery. But as the Canadian economist and famous humorist Stephen Leacock once stated, "I am a great believer in luck, and I find that the harder I work the more I have of it." Kelner noted that the higher the initial dose of UV radiation the greater the magnitude of the subsequent light-dependent recovery. Indeed, systematic examination of the UV radiation dose relationship allowed him to document extremely high levels of photoreactivation, approaching five orders of magnitude at the maximum doses employed. The survival of most UV-irradiated cells generally bears a logarithmic relationship to the dose of radiation. Hence, when survival is plotted on a logarithmic scale and the radiation dose on a normal linear scale, a (mainly) linear relationship is observed. But the typical UV survival curve for *Streptomyces griseus* spores is highly concave in shape, that is, very slight increases in the UV dose result in a dramatic increase in killing in the higher dose range. The doses of UV radiation that Kelner employed yielded extensive killing and hence very large recovery sectors due to photoreactivation. Had he used a more radiation-resistant organism he might not have been as convinced of the photoreactivation effect.

Kelner quickly reproduced his experiments on *Streptomyces* spores with several other microorganisms, including *E. coli*. These results were reported at the Research Conference for Biology and Medicine of the Atomic Energy Commission held at the Oak Ridge National Laboratories in 1949, the same meeting that Dulbecco (and Luria) attended. However, the proceedings of this meeting were not published until 1952 (apparently because some of the speakers were tardy in delivering manuscripts). Parenthetically, Kelner noted in his paper published in these proceedings that in contrast to the results obtained with *Streptomyces*, when UV-irradiated *E. coli* cells were kept in buffer or water in the dark, that is, in the absence of photoreactivating wavelengths of light "the increase in survival rate was about two- to three-fold." Since these cells were scrupulously shielded from photoreactivating light, this recovery phenomenon, small though it was, clearly reflected a light-*independent* repair mode that we now know to be excision repair (see Chapter 3). But Kelner was so absorbed with the more dramatic light-dependent effect that he missed the opportunity of discovering this form of DNA repair. So he failed to pursue his even more interesting observation that "when ultraviolet-irradiated *E. coli* cells are put into a favorable medium at 37°C in the dark, they lose in three hours their ability to recover upon subsequent illumination." What happened of course was that during this extended period of incubation under optimal physiological conditions, the cells removed most of the pyrimidine dimers (the specific substrate in DNA for photoreactivation) by excision repair. Hence the loss of photoreactivation. Kelner provocatively concluded from this observation that "this loss in recoverability suggests that the study of the changes that go on during this three-hour period may offer information as to the cell compounds concerned in photoreactivation." But he did not recognize the operation of a competing DNA repair process.

Dulbecco and Luria also presented papers at the Oak Ridge meeting in April, 1949. Luria on multiplicity reactivation and Dulbecco on photoreactivation. In their biography of Max Delbrück entitled *Thinking About Science: Max Delbrück and the Origins of Molecular Biology*, Ernst P. Fischer and Carol Lipson recounted that when introducing the Oak Ridge symposium, Delbrück made reference to the utility of what he called "the principle of limited sloppiness" with respect to the discovery of photoreactivation. "If you are too sloppy," he said, "then you never get reproducible results, and then you never can draw any conclusions; but if you are just a little sloppy, then when you see something startling you . . . nail it down." Delbrück had previously expressed his frank amazement that photoreactivation had not been discovered earlier. In late 1948, he wrote to Luria that:

> Photoreactivation is a shocker, and it is a miracle that it was not discovered before. It shows that everybody else was working too sloppily to notice it, and you ... too precisely to encounter it. It is the old story of the principle of measured sloppiness that leads to discovery.

Several months following the appearance of Dulbecco's paper, Aaron Novick and Leo Szilard confirmed the results of both Kelner and Dulbecco. Leo Szilard is perhaps best known as the physicist who in the nineteen-thirties first recognized that sustained nuclear fission could provide an enormously devastating explosion, and who drafted the letter to President Franklin D. Roosevelt (co-signed by Albert Einstein) urging the United States government to build the atomic bomb. After the war, Szilard turned his interests to biology, and like many physicists engaged in the Manhattan project he became a vigorous opponent of the militaristic use of nuclear power. He and his colleague Aaron Novick made distinguished contributions to molecular biology in the area of gene regulation. In addition to their confirmation of photoreactivation, Szilard and Novick, like Kelner, flirted with the discovery of a light-independent repair mode. Prior to the appearance of Kelner's 1952 paper, they documented that when UV-irradiated *E. coli* were incubated in the dark they underwent recovery of viability.

> This dark reaction is easily observed if the bacterial suspension is irradiated with ultra-violet and then incubated in the dark at 37°C, for instance for three hours....
> ... the number of survivors after light-reactivation is ... found to be much lower than is obtained when no dark incubation is interposed between ultra-violet inactivation and light-reactivation.

Once again we have early evidence for competing light-dependent and light-independent DNA repair that was not pursued further. Most significantly perhaps was Novick and Szilard's observation that in addition to enhanced light-dependent survival of *E. coli*, the number of *mutants* decreased as a result of photoreactivation.

> At this time we can only say that our results to date are entirely consistent with the view that the effect of light-reactivation on the appearance of mutants among the progeny of the ultra-violet irradiated bacteria is the same as its effect on the number of survivors and that this effect consists in the reduction of the effectiveness of the ultra-violet dose.... This makes it possible to surmise that in our experiments the killing of the bacteria and the production of the mutants might be due to the same chemical effect produced by the ultra-violet irradiation.

It would take another 11 years before the precise nature of the "chemical effect" they were referring to (the formation of pyrimidine dimers) was discovered.

Following the publications by Kelner and Dulbecco, it was by no means evident that photoreactivation was a DNA repair process. As related by Kelner to Rupert, the phenomenon was independently observed by Magni at the University of Pavia, Italy, who suggested that the postirradiation exposure to light destroyed "cellular poisons" that were induced by the UV radiation. Several historians of photoreactivation have pointed out that the "cellular poison" theory of UV radiation and its reversal by some sort of light-dependent process remained prevalent for some time. Dulbecco showed that photoreactivation of bacteriophage occurred if the phage were UV irradiated *before* they infected the bacterial host. This was an important observation which demonstrated that photoreactivation did not require that the cells themselves, the proposed source of the suggested "poisons," be subjected to UV radiation. But one could still imagine that a poison might have been produced in the protein moiety of the phage and that after phage infection this poison was somehow released. Photoreactivation could then work on the poison. Novick and Szilard also used the term poison in their speculations about the effects of UV radiation and the mitigating consequences of photoreactivating light on phage mutagenesis, though a careful reading of their paper suggests that their use of this term was more metaphorical than literal:

> If we are right in assuming that a poison is produced by the ultra-violet rays employed in our experiments, that the amount of this poison is reduced by light-reactivation, and that the amount of poison which is present in the bacteria when they are allowed to multiply determines the number of survivors, it then seems possible that this same "poison" might also determine the number of mutants—resistant to one of the phages—that appears among the progeny of the survivors.

By the early nineteen-fifties, Kelner had extensively characterized the phenomenon of photoreactivation. He astutely honed in on the observation that one of the earliest photoreactivable consequences of UV radiation exposure in bacteria was inhibition of DNA synthesis, and he therefore concluded that "the first consequence of ultraviolet absorption must be a change in nucleic acid molecules." By comparing his results with bacteria to those of Dulbecco with bacteriophage, he hypothesized that:

> The ultraviolet-induced inhibition of desoxyribonucleic acid synthesis is correlated with a general change in the nucleus which results in an inhibition of all or many of the reactions of the cell which are governed by the nucleus—that is, ultraviolet paralyzes nuclear function. Reactivating light removes the paralysis and renders the nucleus functional.

History might be generous enough to concede that such language is as explicit an assertion of DNA damage and repair as "damn it" is to swearing. But the hard facts are that the notion of DNA repair was not categorically promulgated, and neither Kelner nor anyone else exalted the phenomenon of photoreactivation to this status at this time. In recent discussions with John Cairns, whose seminal contributions to the DNA repair field are recounted in a later chapter, he commented that:

> I can remember feeling that photoreactivation was not a real proof that DNA lesions can be repaired. My thinking was that if one form of light can make a lesion, it is perhaps not surprising that another form of light can undo the lesion.

A fundamental conceptual limitation to understanding photoreactivation was the lack of definitive evidence that the target for photoreactivation was indeed DNA. By the late nineteen-forties, the correlation between Avery's transforming principle and DNA was well accepted, and in retrospect the experimental demonstration that UV-irradiated transforming DNA could be photoreactivated in *pneumococcus* would have provided compelling evidence that the primary cellular target for UV radiation was DNA and not protein. But when photoreactivation was discovered, the only known transformable bacteria were *Diplococcus pneumoniae*, the organism used by Avery and his colleagues, and *Haemophilus influenza*. By some confounding anomaly of evolution, it turns out that neither of these organisms are endowed with photoreactivating enzyme!

By 1950, the intellectual notion that cells had evolved specific mechanisms for correcting damage to DNA had not yet penetrated the rapidly evolving world of molecular genetics. Nonetheless, it is noteworthy that in 1952, when discussing the possible mechanism of multiplicity reactivation in light of the recently discovered phenomenon of photoreactivation by Kelner and by his student Dulbecco, Luria used the word "repair." This may indeed constitute the earliest documented use of that term with reference to physiological responses to genomic injury: "The occurrence of photoreactivation... without contradicting the hypothesis of localized damage in discrete determinants, suggested the need for caution in interpreting multiplicity reactivation, since physiological mechanisms of *repair* [my italics] may be involved."

As the half century drew to a close, photoreactivation was widely confirmed in many microorganisms as well as in some higher eukaryotes. Kelner defined the action spectra for photoreactivating light in *E. coli* and in *Streptomyces griseus*, his favorite experimental tool, and correctly postulated that the absorption of light must be effected by specific light-accepting moieties in cells. Very soon after his independent discovery of photoreactivation, Dulbecco elaborated his own studies on photoreactivation of phage. In the course of these experiments, he attempted to reproduce the process in cell-free extracts of *E. coli*. However, since he had no notion that the photoreactivable lesions produced by UV radiation were in the DNA of the phage particles, he incubated these extracts with *intact* UV-irradiated phage particles, and naturally failed to observe any effect. Of course, in 1953 the two classic papers in *Nature* by Watson and Crick changed everything. Though it remained to be shown precisely how DNA functioned as the genetic material, the chemical nature of the gene was now revealed and could be thoughtfully examined for potential substrates for the repair of UV and ionizing radiation damage, and even for substrates that might arise spontaneously. A tantalizing reference to the latter category of base damage (and its potential repair) can be traced to the second of Watson and Crick's famous *Nature* letters in 1953 and provides an interesting diversion from the immediate topic at hand.

In what might be cited as another example of extraordinary understatement (the first being their comment that it had not escaped their attention that their proposed structure of DNA suggested an obvious mechanism for its replication), Watson and Crick wrote that:

> Our model suggests possible explanations for a number of other phenomena. For example, spontaneous mutation may be due to a base occasionally occurring in one of its less likely tautomeric forms.

This prophetic insight into a molecular mechanism for spontaneous mutagenesis arose from their extensive familiarity with the phenomenon of tautomeric shifts in the bases of DNA, since this issue had contributed vexing uncertainties to their early attempts at constructing models of the structure of DNA. For a long time, the difficulties of understanding the appropriate geometry of the tautomeric forms was, according to Judson, "the basis for Crick's prejudice against hydrogen bonds to knit up the chains, and for Watson's against putting the sugar-phosphate backbones on the outside." In their *Nature* paper quoted above, Watson and Crick did not provide any quantitative guesses of what "occasionally" might mean. Hence, they did not concern themselves with the catastrophic implications of too many mutations arising during DNA replication and were not motivated to offer a solution to this dilemma. But in his book published in 1988 entitled *What Mad Pursuit: A Personal View of Scientific Discovery*, Crick returned to this dilemma, albeit in a very different context. In reflecting in a philosophical sense about the pitfalls that can plague the theoretical scientist Crick wrote of the "deductive quicksand" that awaits the theoretician who is ignorant of a minor biological process (in this case DNA repair) which relates to a more general (major) biological process (in this case DNA replication), and therefore infers "that a postulated mechanism for the major process could not work."

> Consider, for example, the rate of making errors in DNA replication. . . . . Human DNA has about three billion base pairs (per haploid set) and although we now know that only a fraction of these have to be replicated accurately, the error rate cannot be greater than about one in a hundred million (speaking very roughly) or the organism would be torpedoed in evolution by its own errors. Yet there is a natural rate for making replication errors [due to the tautomeric nature of the bases] that it would be difficult to reduce to below about one in ten thousand. Surely, then, DNA cannot be the genetic material since its replication would produce too many errors.

The essential message that Crick wished to convey is that ignorance of the ability of cells to repair replication errors could lead the unwary theoretician to the erroneous conclusion that DNA cannot be the genetic material because it is biologically unsuited as a template for copying itself. "Fortunately," he stated, "we never took this argument seriously." More intriguingly, Crick went on to write that:

> The obvious way out is to assume that the cell has evolved error-correcting mechanisms. . . . Leslie Orgel and I actually wrote a private letter to Arthur Kornberg, pointing this out and predicting that the enzyme he was studying that replicated DNA in the test tube (the so-called Kornberg enzyme) should contain within itself an error-correcting device, as indeed it does.

Back to the history of the photoreactivation story. The ultimate demonstration that photoreactivation really is a DNA repair process catalyzed by a specific enzyme with a strict requirement for visible light came from experiments performed by Solomon Goodgal and Stan Rupert. I have known Stan for many years and was delighted to discover that he resides in Dallas and hence could conveniently visit with me at The University of Texas (UT) Southwestern Medical School. Stan is a native Californian who spent many years on the faculty of the neighboring, but completely independent, UT campus called The University of Texas at Dallas, before recently assuming his emeritus position there. We joked about the fact that in the years before UT Southwestern Medical School acquired its crop of Nobel Laureates, many of the folks associated with the UT Dallas campus were offended by the frequent confusion between it and the UT Health Sciences Center at Dallas (as Southwestern Medical School was then called). To add to the considerable nomenclatural confusion of who worked where in Dallas (much of which derived from the decision to rename the UT Health Sciences Center at Dallas as UT Southwestern Medical Center at Dallas in the late nineteen-eighties), UT Dallas grew out of a private research institute originally called the Southwest Institute for Advanced Studies. The Southwest Institute, founded in 1948, comprised (among other molecular biologists recruited there) a formidable group of photobiologists, including Rupert, the noted photobiologist John Jagger, and Walter and Helga Harm, all of whom made fundamental contributions to photobiology in general and to our detailed understanding of photoreactivation in particular. The Southwest Institute/UT Dallas group was perhaps the beginning of what has more recently been referred to as the "Texas Mafia" in DNA repair, with prominent research groups now established in Dallas, San Antonio, Smithville, Houston, and Galveston. Though our individual research interests never fully intersected, Stan Rupert and I attended and participated in many of the same scientific meetings, and I readily recall his sharp and incisive intelligence projected with his characteristic wry humor through his deep sonorous voice.

The first definitive demonstration of photoreactivation in vitro, and hence the first clear indication for an enzyme-catalyzed DNA repair reaction was published in 1956 based on studies that Rupert carried out in collaboration with Sol Goodgal in the laboratory of Roger Herriott at Johns Hopkins University. Sol Goodgal obtained his Ph.D. under the tutelage of Carl Swanson at Johns Hopkins University. Part of his thesis work involved the study of UV- and X-ray inactivation of *Neurospora conidia* in the late nineteen-forties. His friend Clem Markert was at Caltech at the time and through conversations with Markert, Goodgal heard vague rumors about the reactivating effects of visible light on UV-inactivated cells. One presumes that these emanated from the extensive grapevine to which Delbrück (then at Caltech) was firmly rooted. Goodgal attempted similar experiments and was able to confirm unambiguously the phenomenon of photoreativation prior to reading the published accounts by Kelner and Dulbecco. He explored the energy activation temperature dependence of the process and his training in biochemistry as a graduate student was sufficient to convince him that photoreactivation was enzyme-dependent. "It was clear to me," he told me, "that there was an enzymatic reaction going on, but of course I had no idea what it might be."

Upon completing his graduate studies and a stint in the Department of Sanitary Engineering at Hopkins (where, I was relieved to hear, his efforts were confined to work with isotopes rather than toilets), he joined Roger Herriot's laboratory as a postdoctoral fellow in the Department of Biochemistry in the School of Public Health and Hygiene. Herriot was an accomplished phage geneticist who had carried out elegant electron microscopic studies on the morphology of the T series of phages. According to Judson, Herriott was possibly the first to recognize that phages may infect bacteria by injecting their DNA rather than by entering the cells as entire phage particles, as was generally assumed. Judson related part of a letter that Herriott wrote to Al Hershey about this in late 1951:

> I've been thinking—and perhaps you have, too—that the virus may act like a little hypodermic needle full of transforming principles; that the virus as such never enters the cell; that only the tail contacts the host and perhaps enzymatically cuts a small hole through the outer membrane and then the nucleic acid of the virus head flows into the cell.

By 1952, Al Hershey and Martha Chase had carried out their classic Waring blender experiments proving Herriot's suggestion. (The interested reader is referred to Judson's book for a full accounting of these.) In these experiments, Hershey and Chase differentially tagged the phage proteins and DNA with radiolabeled sulfur and phosphorus, respectively. They infected cells with the radiolabeled phage and showed that if the mixtures were vigorously agitated in a kitchen blender (which dislodged the empty phage ghosts from the surface of the cells more effectively than any recognized laboratory device), only the labeled phosphorus entered the cell. This result, coupled with Dulbecco's earlier demonstration that photoreactivation of UV-irradiated phage could take place only *after* the phage infected bacterial cells, pointed to the phage genome as the specific target for UV radiation damage and for the reactivation process. However, this remained to be formally proven.

While in the Biochemistry Department in the School of Public Health and Hygiene (where he subsequently joined the faculty), Goodgal developed his life-long interest in DNA transformation. He also befriended a then graduate student, I. Robert Lehman. After completing his graduate studies, Lehman left Johns Hopkins to take up a postdoctoral fellowship with Arthur Kornberg in the Department of Biochemistry at Washington University, St. Louis, where he played a pivotal role in demonstrating replication of DNA in vitro and in the isolation of the first DNA polymerase from *E. coli*. Goodgal and Lehman maintained their friendship after Lehman's departure from Hopkins and at some point their discussions centered on the notion of attempting to synthesize transforming DNA in vitro. "So I went out to St. Louis," recounted Goodgal. "We carried out some experiments, which didn't work. But, more importantly, I saw how Bob made *E. coli* extracts, and when I got back to Baltimore I thought to myself why not use the transformation assay to see whether similarly prepared extracts of *E. coli* could photoreactivate UV-irradiated DNA?" By that time, the Herriot laboratory had been joined by young Stan Rupert, who turned out to be an eager and willing pair of hands to help test Goodgal's ideas about the enzymatic nature of photoreactivation.

Rupert obtained his Ph.D. in physics from Johns Hopkins in 1951 and remained there, initially in the physics department and then as an American Cancer Society postdoctoral fellow in the newly formed Department of Biophysics. When I commented on the large number of early molecular biologists who were reformed physicists, he retorted with a twinkle in his eye, "Yes, just a golden time for that to happen because you didn't have to know anything—you didn't have to know any biology to suddenly stumble into this empty territory." Rupert related that following the Watson and Crick papers, "everyone at Johns Hopkins was kind of intrigued" and talk of DNA was on everybody's lips. He too was intrigued about DNA, especially since his work in the biophysics department with infrared microspectrometry was not going anywhere. "As a matter of fact," he told me, "I was a little in the same position as Kelner. What I was doing on infrared microspectrometry wasn't looking very exciting and I was searching for something more interesting." Rupert was aware that Herriot offered a short course in phage and bacterial genetics. So he "locked up the physics lab a couple of days a week and trotted down there to see what I could learn, because I needed an awful lot of coaching in biology."

Rupert learned that Herriot and Goodgal were expert in the transformation of *Haemophilus influenzae* with DNA, because *Haemophilus* had a lot of advantages over *Pneumococcus* in the preparation of transformation-competent cells. "They used radiolabeled DNA to monitor DNA uptake and were able to accurately evaluate many of the experimental parameters that influenced transformation-competence in this organism," he recalled. "They also had an excellent genetic marker, the streptomycin resistance gene, to quantitate transformation." So, a biological assay to measure the functional activity of DNA was readily at hand in Herriot's laboratory. Another element that was critical to testing photoreactivation directly on DNA was Herriot's recently acquired interest in radiation effects. This was in the early days of the Atomic Energy Commission's massive financial commitment to radiation biology and so "anything called radiation, whether it was ionizing or nonionizing, had research support," Rupert told me.

Rupert recalled numerous informal discussions in Herriot's laboratory with Goodgal and the emerging notion that it might be informative to use the transformation assay to prove that the substrate for photoreactivation was indeed DNA. The initial idea was to attempt an in vivo experiment. "The plan was to irradiate the DNA with UV light to inactivate it and show that you don't get as many transformants as with unirradiated DNA. Then we planned to let the cells take up the inactivated DNA by transformation, expose them to photoreactivating light and see what happens," commented Rupert.

When the molecular genetics course was over, Rupert committed one or two, and later more, afternoons a week to Herriot's laboratory, where under Goodgal's patient guidance he learned how to do DNA transformation of *Haemophilus influenzae*. But after several months of such experiments, during which photoreactivation in *E. coli* in vivo was used as a positive control, Rupert and Goodgal (correctly) concluded that *Haemophilus* did not carry out photoreactivation, either following transformation with UV-irradiated DNA or in vivo. Goodgal suggested that since photoreactivation clearly worked in *E. coli*, they should perhaps try in vitro experiments with extracts of *E. coli* and use the *Haemophilus* transformation system exclusively as an assay for photoreactivation in the *E. coli* extracts. "Goodgal made it sound awfully easy," Rupert told me. "According to him, all we had to do was grind up some *E. coli*, centrifuge out the junk, and mix the juice with the UV-irradiated DNA. Then expose the mixture to light, transform *Haemophilus* cells and see whether we got more streptomycin-resistant cells!" Having never done a biochemical experiment in his life, Rupert initially protested that this all sounded much too complicated. But with Goodgal's guiding hand he helped set up an experiment in which UV-irradiated transforming DNA was incubated with an *E. coli* extract in the presence or absence of visible light. They added ATP and lots of magnesium to the extract because Lehman and Kornberg had done so in their experiments with DNA replication. Here's how Rupert described the outcome of the first such experiment.

> I put this thing together and Sol looked in every now and again and we put the plates away on a Saturday night. On the following Sunday morning, he was down in the lab, and he called me at home and said that one of the plates had ten times as many colonies as the rest of them and wanted to know which one that was. So I looked in my book that I had with me and told him that was the one in which the

DNA had been exposed to light. On Monday morning, I trotted down to the lab, and Goodgal asked me where all my stuff was. I pointed to the freezer and showed him where everything was and he told me to give him my lab notebook with instructions as to how I did the experiment, and to get out of there and not to touch anything until he was through. "I'm going to do it all over again myself," he said. He did, and it turned out the same way.

A week after the experiments were completed (June 19, 1956), a symposium on *The Chemical Basis of Heredity* was convened at the McCollum-Pratt Institute of the Johns Hopkins University. I'll comment on this important meeting in a different context in the next chapter. Goodgal was squeezed into the program at the eleventh hour to present his and Rupert's preliminary findings. "I think it took a little while for the discovery to sink in," he told me. "I was sitting next to Cy [Cyril] Levinthal at the meeting, and when I sat down after my talk, he gave me all sorts of accolades. He certainly recognized the importance of our findings immediately." Goodgal and Rupert's results on photoreactivation in vitro were published in the conference proceedings under the title "Photoreactivation of *Haemophilus influenzae* Transforming Factor for Streptomycin Resistance by an Extract of *Escherichia coli* B," the first documented evidence that photoreactivation of DNA was an enzyme-catalyzed reaction.

"I went on with the transformation process," Goodgal told me. But enzymatic photoreactivation became Rupert's life work as a scientist. In subsequent detailed experiments, the enzymatic nature of the restoration of the transforming activity of UV-irradiated DNA was supported by showing that the activity was lost by heating the extracts. Most significantly, Rupert observed that the activity was also lost if the extracts were dialyzed, but could be restored by mixing the "large molecular" nondialyzable and the "small molecular" dialyzable fractions. He therefore correctly suggested that dialysis might separate an apoenzyme from one or more low-molecular-weight chromophores required for photoreactivation. The concluding sentence of a paper published in 1958 by Rupert and his colleagues was as crisp and prophetic as any in the scientific literature before or since: "The problem of photoreactivation has thus become a problem in enzymology and photochemistry to be attacked outside the organization and complexity of an intact cell."

What Rupert and his colleagues could not foresee was that each *E. coli* cell contains only 10–20 molecules of photoreactivating enzyme. Hence, purification of the enzyme from this source was a biochemical nightmare and was not achieved for another 27 years when the advent of recombinant DNA technology facilitated the cloning and overexpression of the phr^+ gene that encodes the photoreactivating enzyme in *E. coli*. But the studies initiated by Albert Kelner and Renato Dulbecco and brought to full flower by Stan Rupert and Sol Goodgal conclusively established that UV radiation causes damage (primarily) to DNA, not to proteins, and that this damage can be enzymatically repaired. The field of DNA repair was born at last and the baby was not at all unattractive to the family of molecular biologists! On the 21st anniversary of their publication, Francis Crick revisited the two famous *Nature* papers that he wrote with Jim Watson and conceded that:

> We totally missed the possible role of enzymes in repair, although, due to Claud Rupert's early very elegant work on photoreactivation, I later came to realize that

DNA is so precious that probably many distinct repair mechanisms would exist. Nowadays one could hardly discuss mutation without considering repair at the same time.

Any remaining reservations about the biological relevance of the enzyme activity detected in extracts of *E. coli* were laid to rest by the isolation in 1962 of a mutant of *E. coli* that was unable to carry out photoreactivation in vivo and was found to be defective for the activity in vitro. The power of mutant organisms at every level of biological organization in providing the ultimate validation of biochemical and molecular biological experiments can never be overstated. As Horace Judson remarked:

> Mutation has been essential to Mendelism at every step, not only as the source of heritable variations—round peas or wrinkled, vermillion eyes or the wild type, rough plaque or smooth—but as the tool for understanding. . . . Mutation fairly launched molecular biology, when Salvador Luria and Max Delbrück in the United States, Jacques Monod in wartime France, devised proofs that microorganisms mutate and therefore have genetics, when George Beadle and Edward Tatum proposed, on mutational evidence, that one gene makes one enzyme, and when Linus Pauling realized that sickle-cell anemia is an inherited flaw in the hemoglobin molecule.

In the early nineteen-fifties, Seymour Benzer, another biophysicist whose considerable scientific talents richly benefited the early years of molecular biology, set out to map the fine structure of the *rII* gene of phage T2 at the nucleotide level by mutational analysis, with the goal of "[driving] formal genetic analysis . . . down to the level of the chemical gene." Armed with an increasingly more detailed map of the *rII* gene, Benzer, in discussions with Sydney Brenner, recognized that a demonstration of perfect colinearity of mutations in a gene and in its corresponding polypeptide would provide conclusive evidence of the instructional relationship between genes and proteins. The potential power of such mutational analyses provided strong incentives for devising ways of generating mutants more rapidly than nature offered spontaneously, and of understanding the molecular basis of their origin. By the late nineteen-fifties, Ernest Freese had pretty much deciphered the molecular basis of mutations caused by base analogs such as bromouracil, and by the chemical nitrous acid, and had defined the essential features of and differences between transition and transversion mutations.

The first UV-radiation-sensitive mutant of *E. coli* was isolated by Ruth Hill in 1958 who, curiously, used UV light both as a mutagen and as a method of mutant selection. She irradiated a culture of about a million *E. coli* cells of a strain called *E. coli* B, plated the cells on agar, and exposed them to UV radiation. Hill was intent on isolating mutants that were more *resistant* to UV light. But of the 22 colonies that survived this mutagenic treatment one (that she called strain B_s) was clearly and reproducibly more *sensitive* to this agent. The mutant isolated by Hill was shown by her to be fully capable of photoreactivating UV-irradiated phage T1. However, the mutant was unable to support the recovery of UV-irradiated phage in the absence of photoreactivating light. This strain later played a crucial role in establishing the biology of so-called "dark (photoreactivating light-*independent*) repair" of DNA, now called excision repair (see Chapter 3). Shortly after this work was published, a second

UV-sensitive mutant of *E. coli* was isolated by Walter Harm. This one proved to be defective in photoreactivation.

Before Walter Harm and his wife and colleague Helga joined the Southwest Center for Advanced Studies in Dallas, he cultivated his interest in DNA repair in his native Germany, where he received his Ph.D. degree in 1951. In correspondence with Harm in 1996, he told me that in the early nineteen-fifties:

> American publications were difficult to come by regularly and it was not easy to find out what had been done and what not. Being basically interested in bacterial genetics (a field that was essentially new in postwar Germany) I entered into a close collaboration with a physicist (Werner Stein) who was about to begin studies of ultraviolet effects in bacterial cells because this didn't require much sophisticated physical equipment, for which money was lacking. We were working on the influence of postirradiation incubation temperature on bacterial survival and investigated a number of other postirradiation treatments by which survival was affected. It became increasingly obvious to us that the fate of the cells was not unambiguously determined by UV irradiation alone. We learned belatedly that similar findings were obtained by researchers in the United States, some of which were published years before our studies. In any case, the few people working in this area at that time (including W. Stein and myself) did not talk about "repair," but rather about "reactivation," "restoration," "recovery," etc., meaning that at the time of irradiation the primary effects (which for several reasons were believed correctly to be in DNA) do not alone determine the fate of the cell.

His limited access to the literature notwithstanding, Walter Harm closely monitored events in photobiology that were rapidly unfolding across the Atlantic. He was fully aware of Rupert's exciting demonstration of photoreactivation in vitro by what he (Harm) referred to as "an enzyme-like substance (PhR-enzyme)." Following a period of fellowship training with Max Delbrück at Caltech in the late nineteen-fifties, Harm returned to Germany and became a member of a group of young molecular biologists established by Max Delbrück at the new Institut für Genetik in Cologne. Indeed, Delbrück took a two-year leave of absence from Caltech in order to oversee the development of this new Institute, which he wished to fashion in the image of the American vertical academic system rather than the kind of pyramidal academic structure practiced at the University of Cologne. This "modernization" of German science that Delbrück wanted to achieve represents a most interesting period in his life and in the evolution of molecular genetics in postwar Germany and merits brief discussion.

Having heard a series of seminars from Delbrück during Delbrück's visit to Germany in 1954, Joseph Straub of the University of Cologne "became convinced that the new genetics represented the scientific future and that it should be carried on at a German university as soon as possible." Following a series of protracted negotiations, Delbrück accepted the directorship of this new facility with clearly stipulated conditions. In 1956, he informed George Beadle, the Chair of Biology at Caltech in writing that: "I said I would be interested if something could be created which could serve as a model for other universities in Germany and in other countries for breaking down the organizational deadlock in which biology finds itself all over the world, with Caltech almost an unique exception." Delbrück assembled a group of faculty consisting of some of the most promising young German scientists, who were given the opportunity to grow and mature scientifically in a fully independent manner. This group included Carsten Bresch, Peter Starlinger, Heinz Zachau, and Walter Harm. Lawrence Grossman, a renowned contributor to the DNA repair field (see Chapter 4), spent a mini-sabbatical with Delbrück in Cologne and recalls that "their political vulnerability to encroachment by senior faculty of the university was protected by Max whose stature in science and in postwar Germany was unswerving." Fischer and Lipson wrote that:

> Max oversaw the construction of an institute designed to promote collaboration. Twenty years before that, such an interaction would have been the exception. Each professor would have jealously protected his own special field, preventing others from interfering. Interdepartmental cooperation was an alien concept for German faculty; eventually faculty members such as Bresch and Harm became convinced of Max's conception.

Delbrück was most interested in photobiology in general and during one of his trips to Cologne he invited Rupert, who was on sabbatical in Copenhagen, to give a seminar about his photoreactivation work. (It was during this sabbatical leave that Rupert wrote to Kelner and received from him the voluminous documentation about the events that culminated in his discovery of photoreactivation.) Rupert had met Harm previously, and during this visit to Cologne they initiated discussions about working together. Delbrück's best intentions notwithstanding, relations between the young faculty group at the institute and the established senior faculty at the university grew tense with time. Harm told me that:

> Delbrück accepted the directorship of the institute with the clear understanding of coming to Cologne after the building was completed and of staying for only two years. Most of us who knew Max and his ties to the United States were sure that Max would indeed only stay for two years because under the existing laws at that time he would lose his U.S. citizenship if he returned to Germany for longer than two years, but the senior faculty and administrators expected him to stay for good.

Delbrück's departure after his promised two-year stay left a huge leadership vacuum, and the institute fell into crisis. Fischer and Lipson relate that in May 1964 Delbrück wrote to the dean from Caltech expressing his disappointment. He stated that:

> I thought we had created a new type of institute for Germany and Europe and hoped that would set an example. The main point for me was to set up a group of people within the university, that could compete with the best modern institutes in teaching molecular genetics and in the corresponding research. In addition to that I wanted to demonstrate ad oculos at a European university the polycephaly [multiple heads] principle as it is known here. It is this principle and not the money that is the true secret weapon of the American universities.

Harm, whose promotion to the rank of professor was one of the unresolved problems that embroiled the institute, left Germany for a faculty position in Roger Herriot's department at Johns Hopkins University, where among other attractions he could profitably renew his association with Stan Rupert. In 1964, he was independently recruited to the Southwest Institute for Advanced Studies in Dallas, and a year later he persuaded Carsten Bresch, head of the genetics division of the Institute, to recruit Rupert there.

While in Cologne, Harm, together with his colleague Brigitte Hillebrandt, set out to isolate a mutant that was defective in photoreactivation. "Brigitte Hillebrandt was my technician," Harm informed me. "Because of the monotonous procedures required for isolating a nonphotoreactivable mutant, I promised her coauthorship of a short publication if she could find such a mutant." Utilizing the conventional technology of nitrous acid-induced mutagenesis, Hillebrandt screened *E. coli* colonies for variants that were unable to support photoreactivation of a strain of phage T4 and found a mutant among the first 100 colonies tested, but no more among the next 300. The focus of his attention was the observation that "under conditions where there is a large *PhR*-effect with *E. coli* B [the wild-type strain used in his studies], no *PhR* at all is found for the *phr*⁻ mutant." A few years later, Rupert confirmed that this mutant was indeed defective in enzymatic photoreactivation by showing that extracts of the mutant failed to restore transforming activity to UV-inactivated DNA in the light. Parenthetically, it is interesting to note that like other students of photoreactivation whose studies preceded his own, Harm documented, but did not elaborate on, the observation that control experiments in the absence of photoreactivating wavelengths hinted at a light-independent recovery effect. "[Photoreactivation on plates] suppresses a dark reactivation effect, which occurs if the irradiated bacteria are kept in liquid suspension for a period of time before being plated."

Shortly after his demonstration of photoreactivating enzyme activity in extracts of *E. coli*, Rupert turned to another source for the enzyme, the eukaryote *Saccharomyces cerevisiae*—common baker's yeast. He found a greater abundance of the enzyme here than in *E. coli*, consistent with contemporary knowledge that yeast cells contain 10–30 times more photoreactivating enzyme than do *E. coli* cells. He also celebrated the fact that his extracts of yeast were not as heavily contaminated with nucleases, which relieved him of some of the vexing problem of nonspecific degradation of the substrate DNA used in the transformation assay. Rupert went on to show that the enzyme present in yeast extracts had similar properties to the enzyme from *E. coli*. By the early nineteen-sixties, Rupert had partially purified the yeast enzyme. He also determined an equilibrium reaction for enzymatic photoreactivation based on his observation that the enzyme bound to UV-irradiated DNA in the absence of photoreactivating light, generating a stable enzyme-substrate complex, which in the presence of light was converted to a product, yielding free enzyme. Despite its abundance relative to the *E. coli* enzyme, purification of substantial amounts of homogeneous yeast photoreactivating enzyme also required gene cloning and overexpression and was also not achieved until the late nineteen-eighties.

As early as 1960, just four years after the discovery of photoreactivating enzyme, Rupert pointed out that "the [remaining] problem of photoreactivation is to determine the chemical nature of the reactivable ultraviolet lesions and the mechanism of enzyme action by which they are repaired in restoring normal functional activity to DNA."

The first of these problems was solved in short order. In the late nineteen-fifties, the noted Hungarian-born biochemist and Nobel Laureate, Albert Szent-Gyorgi, demonstrated that photochemical reactions in aqueous and frozen solutions differ markedly, and suggested that these differences result from ice crystals forcing the solute molecules into solid aggregates. In 1958, the Dutch group of Beukers, Berends, and Ijlstra at the University of Delft in The Netherlands observed that when an aqueous solution of thymine was frozen and then subjected to UV-irradiation, the frozen material exhibited maximal absorption at wavelengths lower than that obtained with unfrozen thymine. This observation suggested that the frozen thymine had undergone a photochemical change. Consistent with the reversibility of other known photochemical reactions involving organic compounds, Beukers and his colleagues showed that reirradiation of a thawed solution of UV-irradiated thymine yielded native thymine. A few years later, they completed a thorough physicochemical analysis of the altered thymine and presented the structure of what is now known as the thymine dimer. This dimerization process involves saturation of the 5'6' double bonds of the two thymine monomers, resulting in loss of the absorption maximum of thymine at 254 nm and the appearance of a new absorption maximum at a lower wavelength.

Similar experiments were independently pursued by Shih Yi Wang at Tufts University, who additionally observed the dimerization of uracil in frozen solution and demonstrated the formation of pyrimidine dimers after the irradiation of thin layers of solid thymine or uracil, consistent with Szent-Gyorgi's hypothesis that in the frozen state these compounds essentially exist out of solution. Wang was also in pursuit of the nature of the DNA damage produced by UV radiation. In a paper published in *Nature* in 1961, around the time that the Dutch workers were pursuing their studies, he wrote:

> I considered that photochemical changes occurring in the solid-state may possibly be related to the irradiation-induced changes of deoxyribonucleic acid (DNA) in biological systems. This concept is based on the Watson and Crick and Wilkins proposal for the molecular structure of DNA. . . . the proximity of the bases may permit the interaction of two adjacent base residues on the same chain to form pyrimidine-pyrimidine, purine-purine or pyrimidine-purine dimers coincident with the breaking of interchain H-bonds.

Subsequent experiments by the Dutch team showed that cytosine as well as thymine dimers form in UV-irradiated DNA. The dimers were shown to be stable to acid hydrolysis and hence could be isolated from a total acid hydrolysate of DNA by conventional paper chromatography. The biological implications of their discovery were not lost to the Dutch group.

> The results described may be of far-reaching importance, for they strongly indicate that the thymine present in deoxyribonucleic acid (and in apurinic acid) is converted in the same remarkable way as thymine itself in frozen solution. Consequently, it is becoming very plausible that the sensitivity of micro-organisms to U.V. radiation is caused from the greater part by the easy conversion of the pyrimidines in the nucleic acids. In view of the essential function of these polymers (in heredity and in protein synthesis) the importance of this sensitivity will be obvious.

In their next sentence, Beukers and his colleagues threw a gauntlet to the world of photobiologists: "However, it will be very difficult to prove this directly." This challenge was eagerly pursued by Richard Setlow and his colleague Jane Setlow, who quickly noted the importance of the chemical characterization of a specific product in DNA after exposure to UV radiation. Setlow had recently been recruited by Alex Hollaender from Yale University to join the scientific staff of the rising biology division of the Oak Ridge National Laboratory. A trained spectroscopist, Setlow did much of his early research on the effects of both ionizing and UV radiation on proteins, DNA, viruses, and bacteria, and was acutely poised to lend the announcement of pyrimidine dimers in DNA to various biological interpretations. Among the most significant of these were his studies on the excision of pyrimidine dimers from DNA, the history of which is detailed in the following chapter.

In 1961, Setlow determined the action spectrum for the splitting of thymine dimers and showed that wavelengths shorter than 254 nm (the absorption maximum for DNA) were most effective. The dimerization and monomerization reactions in pure DNA could thus be easily followed by observing changes in the absorption spectrum of DNA. This simple measurement allowed him to determine the kinetics of dimerization and monomerization as a function of wavelength. He found that wavelengths around 280 nm preferentially resulted in the formation of dimers, whereas wavelengths around 240 nm preferentially split them. No other photoproduct in DNA had such kinetics. This photochemical information led Dick and Jane Setlow to show that transforming DNA could be inactivated by long wavelengths of UV light and then reactivated in terms of its transforming activity by subsequent irradiation with shorter wavelengths. The Setlows also showed that the kinetics of this reaction were identical to those of the monomerization of thymine dimers. Jane Setlow wrote to me in 1995 that "since this phenomenon only involved pyrimidine dimers, such dimers must have been responsible for much of the inactivation of biological activity." "Clearly," Dick Setlow echoed, "dimers were important lesions in DNA!" The phenomenon whereby pyrimidine dimers are monomerized by exposure to short-wavelength radiation came to be called *direct reactivation* to distinguish it from enzyme-catalyzed photoreactivation.

The cherry on top of this fruitful cake was placed by Stan Rupert and his colleague Daniel Wulff who exploited the acid-stability of thymine dimers to show that when DNA was enzymatically photoreactivated in vitro and then subjected to acid hydrolysis dimers were not recovered in the hydrolysate. The indication that both enzyme-catalyzed photoreactivation and short-wavelength direct reactivation operated on the same substrate, suggested very strongly that enzymatic photoreactivation effected the monomerization of pyrimidine dimers.

In her years at Oak Ridge and subsequently at the Brookhaven National Laboratory, where she is still active, Jane Setlow made notable contributions to our understanding of photoreactivation. At a meeting in Chicago in 1965 on *Structural Defects in DNA and Their Repair in Microorganisms*, Stan Rupert introduced Jane as follows:

> Our first speaker has made great contributions to the understanding of this topic. It is probably safe to say that, together with the man who sometimes accompanies her to meetings, she has transformed the study of the subject. Never again will we be satisfied with answers which don't in the end address the explicit chemical processes presumed to underlie radiation phenomena—a thing which seemed quite impossible only a few years ago. Vigorously and ingeniously pursuing a simple working hypothesis to its limit, this lady and her colleagues (along with workers in a few other laboratories) have built up a very comprehensive and coherent account of ultraviolet inactivation and photorecovery. This story will be told to you so

charmingly you will want to believe it all. For this reason she is a very dangerous woman! Her account of these phenomena—as complete as anyone can give you—is certainly much truth mingled with many lies. I can't tell you which is which. And neither can she! But we can be sure that difficulties have been concealed and bypassed, that contradictions have been glossed over, and that reality has been considerably oversimplified.

We know that the medieval church was diligent in seeking out heresy, not only for what it might do to the promulgator, but also for its dangers to the innocent and uninformed. We must be at least as diligent in sifting what this speaker has to say.

I yield the floor to my good friend—until a few moments ago—and very competent competitor, Dr. Jane Setlow.

Jane Setlow made notable contributions to several other aspects of DNA repair, in particular repair in the nonphotoreactivable but transformable bacterium *Haemophilus influenzae*. She was also intrigued by the extraordinary UV radiation resistance of the bacterium *Micrococcus* (now called *Deinococcus*) *radiodurans*, an organism that she referred to as "the Babe Ruth of DNA repair." Her studies on *Deinococcus* paved the way for a contemporary resurgence of interest in this fascinating organism and the demonstration that it is endowed with multiple mechanisms for repairing pyrimidine dimers. Jane also shared with me her reflections on an important sociological issue in science, being a female scientist in a male-dominated world.

When I decided to enter graduate school after being a technician for a while, I had four young children, a sick mother living with us, a 16-room house, and very little money. In addition, I wasn't adequately prepared for biophysics, since I hadn't majored in physics or math in college at Swarthmore. Thus I had to take an undergraduate course in advanced calculus, without having had the prerequisites for that. In those days a female had to get special permission to be in an undergraduate course at Yale, and this took me a long time to achieve. Since I was having difficulty coping with my work as well as all the domestic chores at home, I called the National Institutes of Health, which had given me a predoctoral fellowship, to find out whether I could get some extra cash because of having four children. The written rules at the time said that a male could get extra cash to help support his wife and children no matter how much his wife earned, but a female could only get extra cash if she was the sole source of support of her children. I got to talk to the biggest wheel at NIH, a Dr. Hatchett, who said he couldn't do anything for me, but he was going to try to fix the rules for future female students! Then he said, "Frankly, we have never had a female fellowship holder with so many children as you have."

In later years I was at a meeting of biophysicists (with almost no females present), and ended up going out to dinner with 18 males. They made a great show of opening the door for me, and parading after me to our table. The hostess came up to me and said, "Do you mind if I ask you a personal question?" I said OK, and it was, "What in hell have you got?" I explained about the scientific meeting, and we both laughed.

Much later I finally made it through school, got my degree, and eventually quite a few years later found that in one particular case (the only time it happened), I was able to benefit from the experience of having raised four fractious young ones. I had just been elected to the presidency of the Biophysical Society. The President takes over the Council on the last day of the meeting. It was about 7:00 a.m. and the majority of the members seemed to have massive hangovers. I was very nervous about having to deal with this collection of male distinguished biophysicists. The hotel had goofed badly, and there was no coffee until much later. Finally a small plate of buns was brought in. Much to my amazement a considerable portion of the group began fighting over the buns. This was a situation with which I was *thoroughly* familiar, and my nervousness instantly vanished when I declaimed, "Shut up, sit down, and I will divide the buns." They blinked at me, sat down, and shut up, and the meeting got going.

The second of the two primary issues concerning photoreactivation that Rupert identified in 1960 was the mechanism of enzyme action by which the reactivable ultraviolet lesions are repaired in restoring normal functional activity to DNA. This brings us to the more recent history of photoreactivation. As stated earlier, attempts to purify photoreactivating enzyme from various sources, notably *E. coli* and the yeast *Saccharomyces cerevisiae*, were initially fraught with complexities and artifacts. Very early after he discovered photoreactivating enzyme, Rupert calculated that there was so little of the enzyme constitutively present in *E. coli* and yeast that he quickly (and wisely) made the decision to "leave the purification to others." It was not until 1980 that a photoreactivating enzyme was purified from yeast by Harold Werbin and his colleagues. Denaturation of the purified protein yielded a low-molecular-weight, light-absorbing moiety, flavin adenine dinucleotide, the first tangible evidence of a chromophore. Subsequently, it was established that all photoreactivating enzymes have two chromophores.

Shortly after recombinant DNA technology exploded on the scientific scene, Aziz Sancar (now at the University of North Carolina and a prolific contributor to the biochemistry of DNA repair) cloned the *E. coli phr* gene during his tenure as a graduate student with Rupert at The University of Texas in Dallas. This was the first DNA repair gene to be anointed into the new age of molecular biology. With an energy and intensity well known to those who have worked directly with him, Sancar went on to purify and characterize the *E. coli* enzyme in exquisite detail. Sancar's wife and sometime colleague, Gwendolyn Sancar, achieved similar success with photoreactivating enzyme from yeast, and this class of enzymes has since been purified and characterized from multiple prokaryotic and eukaryotic sources.

When Stan Rupert and I reminisced about the history of the photoreactivation story and the formal birth of the DNA repair field in the spring of 1995, we concluded his visit by stopping off at Johann Deisenhofer's laboratory in the Howard Hughes Medical Institute at The University of Texas Southwestern Medical Center. Deisenhofer, an X-ray crystallographer who received the Nobel Prize in Chemistry in 1988, had recently solved the crystal structure of the *E. coli* photoreactivating enzyme. As he guided us through the computer model of the structure, turning the protein this way and that, and showing us the proposed catalytic site and the contact sites for the two chromophores, I could only marvel at this extraordinary moment in history when I stood in Deisenhofer's darkened computer room flanked on one side by the person who first identified photoreactivating enzyme in crude extracts of *E. coli*, and on the other by the person who a mere 38 years later had brought our understanding of this enzyme to the cutting edge of modern science.

CHAPTER 3

The Emergence of Excision Repair

UNLIKE PHOTOREACTIVATION, which is specifically and uniquely directed to pyrimidine dimers in DNA, but which was evidently lost as a specific DNA repair mode during the evolution of most placental mammals, excision repair is a much more general and ubiquitous process. The discovery of photoreactivation formally secured the concept of DNA repair in nature, but the discovery and subsequent elaboration of the full repertoire of excision repair modes established the generality of repair for a large medley of implied lesions. As the term implies, during excision repair chemically altered nucleotides are enzymatically *excised* from the genome and replaced by normal ones. This theme of nucleotide loss and nucleotide replacement can be effected by an astonishing variety of enzymes, which eliminate an equally astonishing array of abnormal bases; not only those damaged by environmental physical or chemical mutagens or by spontaneous chemical alterations, but also bases that are mispaired as a result of errors during DNA replication, and bases that are inappropriate to the usual chemistry of DNA, such as the presence of uracil instead of thymine.

Nature solved the replacement problem for all excision repair modes by the process of DNA synthesis. Whereas in principle the accurate duplication of genetic information by DNA replication requires only a single-stranded genome, the replacement of nucleotides lost from one DNA strand by their excision has an obligatory requirement for the complementary strand to serve as an informational template. This observation, coupled with the prevalence of UV radiation from the sun as a natural source of base damage, suggests that the theme of nucleotide loss and replacement as a strategy for coping with genomic injury may have provided the primary evolutionary selection for double-stranded DNA. Bob Haynes, Philip Hanawalt, and others in the young DNA repair field subscribed to this evolutionary notion enthusiastically, but the idea was not widely promulgated outside the field. Nonetheless, Haynes recalled that following the celebrated Watson and Crick papers published in 1953, there was some skepticism about their now famous extrapolation from the duplex structure of DNA to a suggested mechanism for its replication. Watson and Crick stated that: "It has not escaped our notice that the specific

[base] pairing we have postulated immediately suggests a possible copying mechanism for the genetic material."

"The complementarity of the DNA strands clearly provided for the mechanism of replication," said Haynes, "but it still, to some people at least, seemed odd that you would have the genetic information represented twice over when you only really need one strand for replication and transcription." Shortly after the formal discovery of excision repair in 1964, Haynes together with Phil Hanawalt, both of whom we shall hear more about later in the chapter, wrote an article for the popular magazine *Scientific American* entitled "The Repair of DNA" in which they noted that:

> The two strands of DNA are complementary because adenine in one strand is always hydrogen-bonded to thymine in the other, and guanine is similarly paired with cytosine. Thus the sequence of bases that constitute the code letters of the cell's genetic message is supplied in redundant form. Redundancy is a familiar stratagem to designers of error-detecting and error-correcting codes. If a portion of one strand of the DNA helix were damaged, the information in that portion could be retrieved from the complementary strand. That is, the cell could use the undamaged strand of DNA as a template for the reconstruction of a damaged segment in the complementary strand.

Aside from its significance as a repair process in its own right, the discovery of photoreactivation provided a powerful intellectual launch pad for the discovery of excision repair. The post-UV radiation recovery phenomenon alluded to by Alex Hollaender and his colleagues in the mid nineteen-thirties and the phenomenon of multiplicity reactivation of T phages documented by Luria in the following decade, began to take on new and important significance as other observations on the recovery and reactivation of UV-irradiated cells and bacteriophage emerged. For example, while working with Delbrück at Caltech in the early nineteen-fifties, the Swiss bacteriologist Jean Weigle reported a new phage reactivation phenomenon, which he called "UV restoration." Weigle observed that when phage λ was exposed to inactivating levels of UV light and then plated on its *E. coli* host, the level of phage survival was increased if the phage-bacterial complexes, or indeed just the host, were also exposed to UV radiation. And just a few years later, Alan Garen and Norton Zinder at the Rockefeller Institute observed that *prior* irradiation of *Salmonella typhimurium* cells with nonlethal doses of UV radiation improved the survival of UV-irradiated phage p22. Influenced by the multiplicity reactivation studies of Luria, Garen and Zinder proposed that this reactivation might be effected by recombination between the host and phage genomes. However, the demonstration by Ruth Hill and her colleagues that the survival of various UV-irradiated T phages was significantly diminished when plated on her UV-sensitive B_{s-1} strain of *E. coli* (see Chapter 2) cast doubt on this explanation, and suggested the more likely explanation that the reactivation of UV-irradiated phage and the recovery of UV-irradiated host cells were dependent on one and the same process, a process that was termed *host cell reactivation*.

Ruth Hill was Professor of Biology at York University in Toronto, Canada when she died suddenly of a cerebral hemorrhage in 1973. In a tribute to her significant contributions to DNA repair written a year later in the proceedings of the 1974 DNA repair conference convened in Squaw Valley, Bob Haynes stated that:

Hill won an important and secure place in the history of biology by virtue of her discovery in 1958 of the first radiation-sensitive mutant bacterium, strain B_{s-1}, of *Escherichia coli*. The isolation of this mutant came as a surprise to radiation geneticists, who up to that time had been more concerned with the isolation of radiation-resistant strains. Furthermore, the extent to which the mutant alleles in B_{s-1} affected sensitivity, especially to ultraviolet light, was vastly greater than had ever been observed before for other physical, chemical, or biological modifiers of radiobiological action. It was not surprising that the very existence of such a mutant should immediately have attracted the interest of radiation biologists, most of whom were accustomed to dealing with three- or four-fold changes in radiosensitivity but certainly not with sensitization factors of 100 or more. The many theories and experiments stimulated by Ruth's discovery of *E. coli* B_{s-1} have, to say the least, changed the complexion of cellular radiobiology. But even more significantly, comparative studies of B_{s-1} and wild-type strains, designed to elucidate the macromolecular basis of wild-type resistance, led to the discovery in 1964 of DNA excision repair, one of the few fundamental processes in molecular biology that was not foreseen by the clairvoyant pioneers of that field.

Other reactivation phenomena gained increasing attention in the nineteen-fifties. Walter Harm provided evidence that the enhanced resistance of bacteriophage T4 to UV radiation compared to its genetically related but distinct cousin phage T2, originally observed by Luria in the nineteen-forties, was determined by a single gene that he suggested might encode a DNA repair enzyme. Harm wrote to me that:

> One of my contributions to the field of repair was the first experimental evidence that a gene of the phage T4 is crucial for the repair of its own DNA. I was intrigued by the fact that the phages T2 and T4, in spite of their similarity in DNA, showed considerably different UV survival curves when infecting the same type of *E. coli* cells. Phage recombination experiments by Luria in 1949 and a more detailed analysis by George Streisinger had shown that the sensitivity difference is controlled by a single locus. My experimental work showed that the lesser UV-sensitivity of T4 was in fact due to repair, for which T2 is deficient. After publication of these results in 1958, 1959, and 1961 I was able to isolate a few T4 mutants, one of which (later termed $T4_{v1}$) showed all the characteristics of T2.

The "single locus" that Harm reasoned must determine the enhanced UV resistance of phage T4 was first called the *u* gene (for *u*ltraviolet radiation) and the phenomenon by which it specified this resistance was termed *u*-gene reactivation. Later nomenclatural clarification, the details of which are recounted in the textbook *DNA Repair* resulted in a change of the term *u*-gene to v^+-gene. I will consider the events that ultimately led to the elucidation of the nature of the DNA repair function specified by the v^+gene of phage T4 later in this chapter and more fully in the next one. Meanwhile, Walter Sauerbier, a colleague of Harm at the Institut für Genetik established by Max Delbrück at the University of Cologne, examined Garen and Zinder's recombination hypothesis of host cell reactivation more closely. Sauerbier reasoned that if host cell reactivation of phage DNA was indeed due to the exchange of genetic material between the host and phage genomes, it should be observed with damage in the phage genome other than that caused by UV radiation. In a series of papers published in the early nineteen-sixties he showed that neither bromouracil substitution of thymine in phage DNA, nor treatment of the phage with nitrous acid was subject to host cell reactivation. Since both photoreactivation and *u*-gene reactivation were apparently UV-radiation-specific recovery phenomena, Sauerbier concluded that: "HCR [host cell reactivation] is caused not by a genetic exchange between phage and bacterial genomes, but by an enzymatic reactivation comparable to photoreactivation and *u*-gene reactivation in T4 and T2*u*."

Another important general insight that emerged during the nineteen-fifties was the realization that the sanctity of the gene could indeed be violated by the metabolic machinery of the cell. DNA polymerase was discovered by Arthur Kornberg and his colleagues at Washington University in St. Louis and the existence of nucleases that degrade DNA was also firmly established. The general concordance of the budding field of DNA biochemistry, the discovery of photoreactivation and the elucidation of its mechanism, the identification of pyrimidine dimers as a specific form of DNA damage, and the emergence of other "UV reactivation" phenomena led scientists interested in understanding cellular responses to radiation damage to look expectantly, and indeed confidently, to the discovery of new forms of DNA repair. Haynes related to me:

> I vividly remember having a discussion about this in my lab with a couple of graduate students and others. This was after we knew about Rupert's work on photoreactivation, but it was before we knew anything about Setlow's interest in excision repair. We stood in front of a blackboard and said, "Let's try and write down other possible ways in which DNA might be repaired."

I'll allude more to this blackboard discussion later. In a similar anticipatory vein, when recalling the period immediately following the discovery of photoreactivating enzyme in the late nineteen-fifties, Walter Harm told me that: "From that time on, the concept of repair of UV-damaged DNA became more popular, and the search was on to find out whether some of the other known 'recovery' effects were perhaps also a consequence of enzymatically controlled repair."

This fertile intellectual environment from which excision repair ultimately emerged confounds the historiography of this topic. The direct experimental discovery of excision repair reflects the parallel efforts of a number of key players, notably Dick Setlow and his colleagues Paul Swenson and Bill Carrier at Oak Ridge National Laboratory, Paul Howard-Flanders and his postdoctoral fellow Dick Boyce at Yale University, and Phil Hanawalt and his graduate student David Pettijohn at Stanford University. Aside from the primary importance of their contributions (particularly those of Setlow and his colleagues) in formalizing the existence of a DNA repair mode that operates independently of photoreactivation, the direct experimental elucidation of excision repair demystified the many vague allusions to alternative repair modes generally couched in terms such as "recovery" and "reactivation," and placed this repair paradigm firmly in the arena of modern molecular biology. In a letter written to me in 1995, Bernard Strauss succinctly evaluated his view of the primary historical contribution of the efforts of Setlow, Howard-Flanders, and Hanawalt as follows:

> I think it necessary to distinguish between what the radiobiologists thought and what the molecular biology community paid attention to. The radiation biologists did killing curves and tried to understand what was going on. I believe that until there was some biochemistry attached to these curves none of the molecular biol-

ogists engaged in working out the details of the genetic code and of transcription paid the slightest attention. What Dick Setlow accomplished was to attract the attention of mainline biologists to what was going on.

In a similar vein, Matthew Meselson recounted to me that in the early nineteen-sixties "the only radiobiologists able to command some audience outside of the field were Dick Setlow and Alex Hollaender. Many of these people had a very limited knowledge of modern genetics, and that was the key necessary to open the door."

The immediate significance of the experimental elucidation of excision repair of DNA notwithstanding, it must be appreciated that the investigative teams who took center stage in this effort were in retrospect intellectually well poised to explicate this phenomenon because the notion of the repair of DNA damage other than by photoreactivation was clearly in the air, and was circulating freely in the informal grapevine of seminars and meetings, not to mention more formal allusions and speculations in the literature. As we shall presently see, Setlow got there first and got there definitively! But before I recount the specific contributions of these groups to the discovery of excision repair, it is both relevant and informative to visit the history of some of the research efforts and attendant hypotheses that anticipated, and thereby indirectly contributed to, this experimental breakthrough. When I discussed this historical nuance with Dick Setlow, he gracefully ventured that "the atmosphere at the time is just as historically important as the details of the particular experiments."

When the photoreactivation story was emerging during the late nineteen-forties and early nineteen-fifties and there were increasing references in the literature and at scientific meetings to postirradiation recovery phenomena in general, a related but distinct area of investigation was motivated by the quest to understand the mechanism of the induction of mutations in bacterial cells by UV radiation. Early studies led to the recognition that, as was the case with survival, the yield of mutations in cells was influenced by postirradiation conditions and treatments. Based on her investigations on the mechanism by which UV radiation results in the induction of a particular class of mutations in bacteria that she (fortuitously) happened to be studying, Evelyn Witkin was led to the prophetic notion in the early nineteen-sixties that some type of enzyme-catalyzed light-*independent* (dark) repair was one of the variables that influenced their frequency.

Witkin traces her interest in genetics to her time as an undergraduate student at New York University. There she encountered the translated papers of Trofim Lysenko, the Russian plant breeder who rejected Mendelian genetics in favor of the notion that heredity could be altered in desired directions in response to environmental manipulations. Many years later in recounting her professional beginnings, Witkin wrote that:

> My decision to begin graduate work in genetics at Columbia was largely motivated by a desire to test Lysenko's ideas. After a few months of study of real genetics I concluded that he was either a charlatan or an ignoramus, or both. But by then, I had become passionately addicted to thinking about genes and mutations, and I am forever grateful to Lysenko for leading me toward a lifetime of joyful work in genetics.

As was the case with many of the early stalwarts of genetics and molecular biology, Cold Spring Harbor Laboratory figured prominently in Witkin's scientific growth and maturation.

> My interest in ultraviolet mutagenesis in bacteria began . . . when I arrived in Cold Spring Harbor as a graduate student, to learn all I could during that summer about the newly emerging field of bacterial genetics. I had decided (with the encouragement of my advisor at Columbia University, Professor Theodosius Dobzhansky) to begin my predoctoral research on induced mutagenesis using bacteria, instead of *Drosophila*, as I had planned before reading the landmark paper of Luria and Delbrück. On my first day at Cold Spring Harbor in June of 1944, Dr. Demerec handed me a culture of *Escherichia coli* B, pointed to an ultraviolet lamp, and said "Go, induce mutations!" I was terrified. I had never had a course in microbiology, and barely knew how to hold a pipette, let alone how to observe the rudiments of sterile technique.

Witkin was a quick learner. In addition to mastering pipetting and sterile technique, she was soon led to the discovery of a novel mutant of the *E. coli* B strain which displayed *enhanced* resistance to UV radiation; the opposite phenotype of Ruth Hill's UV sensitive B_{s-1} mutant discovered later. Having been directed by Demerec to learn how to induce mutations with UV radiation, Witkin immediately began to set up survival curves with *E. coli* B to get some idea of the appropriate dose ranges to use for her mutagenesis studies. Mercifully, her relative lack of experience was not complicated by the unwitting exposure of her plates to photoreactivating light that caused Albert Kelner so much trouble. Nonetheless, she related that she "had no idea what range of doses to use; there were no published survival curves at that time. I went overboard and used a much too high range of doses the first time. All my plates were blank except for one plate which had four colonies on it." She called one of these B/r (for *r*esistance). This strain attracted significant interest. This observation was made immediately following the building and use of the atomic bomb and Witkin recalls that Leo Szilard, the physicist who was intimately involved with the Manhattan project, was "very intrigued that a single mutation in an organism could enhance radiation resistance by as much as a hundred fold." You will recall from the previous chapter that his humanitarian interest in protecting against the lethal effects of radiation was a strong motivating factor in Kelner's interest in the recovery of cells from the effects of UV light. Indeed, following World War II much of the national research effort in radiation biology was specifically dedicated to discovering ways by which cells, and even whole organisms, might be protected (naturally or by experimental manipulation) from the lethal effects of radiation, especially ionizing radiation.

Witkin's mutant strain *E. coli* B/r enjoyed popular usage in photobiological studies because its response to UV radiation resembled that of normal strains more than that of the then widely used parental strain *E. coli* B. The latter was in fact more sensitive to killing after exposure to UV light and underwent filamentous growth, a state in which the cells failed to divide after replicating their DNA, thereby generating long snake-like structures. The mutation that Witkin generated in the more UV-radiation-resistant B/r strain reverted this phenotype and was eventually mapped to a gene initially called *sul* and later *sfi* (for *s*uppressor of *fi*lamentation). The deciphering of the genetic regulation of filamentation in UV-irradiated *E. coli* is part of another fascinating story told later in the book; the discovery of the so-called SOS phenomenon, a chapter in the history of the discovery of DNA repair mechanisms in which Evelyn Witkin also figured prominently. In her later years, she commented on this extraordinary coincidence:

> The work on filamentous growth had been a distraction, albeit one I had found quite fascinating, and I now put it aside, little dreaming that twenty five years later I would come round to it again as a component of the DNA damage-inducible SOS response of *E. coli*, and would "know the place for the first time."

Following her productive first summer at Cold Spring Harbor, Witkin returned to the Laboratory the following year to complete work for her doctoral thesis and remained there for another ten years, initially as a postdoctoral fellow with Demerec, and later as a member of the scientific staff. In the early nineteen-fifties, she initiated studies on a then poorly understood phenomenon in bacterial mutagenesis referred to as the "delayed effect." Demerec and his colleagues had invented the now widely used auxotroph/prototroph system for selecting mutants. When *E. coli* cells that are defective in a gene which determines a particular nutritional requirement (*auxotrophic* mutants) are plated on medium largely depleted for the nutrient in question, they typically die unless they undergo mutational reversion to a nutrient-independent (*prototrophic*) state. Curiously, when this type of mutational reversion was induced by exposure to UV radiation, the prototrophic mutants appeared to arise continuously for many generations following exposure. It was appreciated that this did not necessarily imply that mutational events continued to be generated during multiple division cycles. The more likely explanation entertained at the time was a delayed phenotypic expression of mutations, hence the term delayed effect. Systematic and diligent experimentation eliminated a number of hypotheses formulated to explain the delayed effect. She recently wrote that:

> Running out of hypotheses concentrates the mind wonderfully, and it forced me to recognize that increasing mutation yields and increasing numbers of post-irradiation cell divisions might well be parallel but independent and experimentally separable consequences of increasing nutrient broth concentrations.

This indeed turned out to be the case. Witkin conclusively showed that there was really no delayed effect at all; the mutational yield was in fact determined by the concentration of nutrient broth in which the bacteria were placed during the first postirradiation hour, a fraction of the time required for the first cell division. In her recent historical reflections, she wrote that:

> Having solved one mystery, we were left with a new and intriguing question. Why was the induced mutation yield so dependent upon the amount of nutrient broth present during the first post-irradiation cell division?

Witkin shifted the focus of her interest to this new problem. She related the requirement for nutrient broth to a requirement for amino acids and ultimately to a requirement for protein synthesis. She discovered that the postirradiation mutational yield was highly dependent on protein synthesis and that experimental perturbations that interfered with protein synthesis resulted in a rapid and irreversible *loss* of most of the prototrophic mutations that she was scoring. This mutational loss came to be called *mutation frequency decline* (MFD). Witkin's detailed studies on the phenomenon of MFD will not be recounted here. The issue germane to the present discussion is that these studies, aided by her interpretation of the results of others, eventually led her to the suggestion that the yield of mutants induced by UV radiation in her experimental system was influenced by some sort of DNA repair process that removed premutational UV-radiation-induced damage under certain conditions. She later recounted that:

> By 1961 [several years before the discovery of excision repair], my own intensive study of the phenomenology of MFD, taken together with the work of [Charles] Doudney and [Felix] Haas [as well as Bryn Bridges and Margaret Lieb] had led me to propose: (1) that UV mutagenesis begins with the production of potentially mutagenic UV damage directly in DNA; (2) that MFD reflects repair of this damage by an unidentified "dark repair" system, probably enzymatic in nature; (3) that "dark repair" is inhibited by post-irradiation conditions favoring protein synthesis, and also by post-treatments with caffeine, acriflavine and other basic dyes that bind DNA; (4) that the yield of UV-induced prototrophs is irreversibly "fixed" by DNA replication, and is proportional to the amount of unrepaired damage remaining in the DNA when replication occurs. MFD was thus seen as "repair before replication," and mutation fixation (MF) as "replication without repair."

In writing about this relationship between mutagenesis and DNA repair in her historical review published in 1994, Witkin was quick to acknowledge that like many of her contemporaries investigating the effects of UV radiation in the late nineteen-fifties and early nineteen-sixties, her thinking was ripe for the notion of a new DNA repair mode.

> Two advances in understanding UV effects had made the idea of enzymatic repair of UV lesions in DNA seem reasonable. Thymine dimers had just been identified as a major UV lesion in DNA, and the photoreversal of both lethal and mutagenic effects of UV irradiation had been shown to depend upon the visible light-dependent action of a photoreactivating enzyme.

Though not explicitly stated in these historical reflections, Witkin recalls that she was led to think and certainly to write about DNA repair as a determinant of prototrophic mutation frequency as early as the mid nineteen-fifties, though not in the precise terms enunciated above. The Cold Spring Harbor Symposium in 1956 celebrated a revisiting of progress in molecular genetics. The symposium was entitled *Genetic Mechanisms: Structure and Function*. Milislav Demerec was still director of Cold Spring Harbor Laboratory, and the list of attendees in the fledgling DNA repair field now included Tikvah Alper, Fred De Serres, Alex Hollaender (of course), Albert Kelner, Margaret Lieb, George Streisinger, and Evelyn Witkin, then at the State University of New York in Brooklyn, New York. Witkin presented a paper on UV-induced mutations in bacteria in which she resoundingly dispelled the myth of the delayed effect and formulated the then budding hypothesis about the role of DNA repair, quaintly identified as metabolic factor "X:"

> One possibility is that X is the process of repair of genetic damage. According to this view, induced prototrophs arise primarily from a fraction of survivors that has undergone genetic damage as a result of irradiation. Protein synthesis favors repair of this damage and if repair is accomplished before the end of the sensitive period (DNA duplication?) the damaged cells survive, with a high probability of surviving

as a mutant. If repair is not accomplished within the critical period, the damage is lethal.

In her retrospective historical review of this investigative period, Witkin reminds us how fortuitous her intellectual linkage of UV-induced mutagenesis to dark repair really was. At the very outset of her studies on the "delayed effect," she specifically sought *E. coli* auxotrophs that would allow her to establish optimal conditions for her experiments on UV mutagenesis. She told me that she very much wanted to work with strains that "would yield a low spontaneous mutation rate, high yield of UV-induced prototrophs, uniform size and appearance of revertant colonies, and mutant colony counts reaching a stable endpoint after just two days of incubation:"

> What I had no way of knowing until years later was that my screening test automatically excluded all auxotrophs except those that owed their nutritional defects to amber or ochre nonsense mutations, and that the UV-induced prototrophs that satisfied my criteria were almost invariably efficient nonsense suppressor mutations, each arising at *a unique site in one of only two linked tRNA genes!* Inadvertently, I had set up a system that focused on the idiosyncratic repair of UV lesions located at one or the other of only two particular bases, of the millions of bases in *E. coli* DNA. But back in 1952, a decade before nonsense suppression via tRNA mutations was known, my well-behaved strains seemed the perfect material with which to study what I then presumed to be the generalizeable reactions of genes to ultraviolet light. In retrospect, it was a case of "too smart is dumb!"

Subsequent studies indeed verified that MFD is a specific response of UV-induced mutations to prototrophy. Other types of mutations, such as those leading to antibiotic resistance, did not exhibit MFD. Insights into the more general relationship between UV radiation damage and other types of mutations that are not subject to MFD emerged years later through Witkin's continued and unswerving dedication to the study of mutation induction by UV light. Nonetheless, history must credit her with the conceptual formulation of a light-independent, enzyme-catalyzed DNA repair process as an explanation for MFD. Witkin was thoroughly convinced that MFD was related to a dark repair mode which removed radiation-induced premutational base damage, and she garnered a wealth of information about some of the general properties of this (hypothetical) repair mode based on her experimental observations on MFD. So much so, that when she visited her friends Dick and Jane Setlow at Oak Ridge in 1964, just before Dick Setlow published his experimental results that firmly demonstrated the existence of excision repair in *E. coli*, she challenged him with a series of predictions about the process. She related:

> I gave him a list. I told him that excision repair is going to be inhibited by caffeine, and that it's going to be inhibited by acriflavine and a whole bunch of basic dyes which bind DNA that I had investigated. I told him that it's going to require an energy source, and that it is going to be UV dose-dependent. I made a bet with him that all of these facts would turn out to be true. Of course he hadn't tested any of them yet. I got a preprint from him later on with a little handwritten note on the cover saying, "You win the bet!"

When I recounted this story to Dick Setlow, he amusedly recalled the loss of this little bet.

Interpretations of the results of experiments of a completely different nature which were executed in the early nineteen-sixties also supported the notion of excision repair of DNA. Like many biologists interested in the effects of radiation on biological systems, Bob Haynes began his formal scientific training as a physicist. He dabbled in theoretical nuclear physics at McGill University in his native Canada and completed his graduate work in biophysics under the mentorship of Alan C. Burton, who was investigating the rheology of human blood flow at the University of Western Ontario. When Haynes naively asked his mentor what rheology was, he was curtly told that he had "best find out soon!" Haynes learned that rheology is in fact the study of the (non-Newtonian) flow properties of particulate suspensions such as blood, a far cry from his recently cultivated interest in genetics. Nonetheless, he acknowledges that in retrospect his training in physical biology, in particular its contribution to his natural theoretical and analytical bent, served him (and the DNA repair field) well. In a recent autobiographical discourse published in a historical volume entitled *The Early Days of Yeast Genetics*, Haynes commented that:

> Like other young physicists of the day, my curiosity was aroused by reading Erwin Schrödinger's (1944) book *What Is Life?*, and also F. Dessauer's (1954) *Quantenbiologie*. I became imbued with the fantasy that sufficiently clever physicists could, through further developments in the quantum mechanics of molecular interactions, solve the deep problems of biological reproduction and heredity. Biochemistry was messy and could be safely ignored. I was by no means alone among physicists in spouting such arrogant nonsense.

Haynes's self-deprecating apology notwithstanding, this view of biochemistry in the late nineteen-forties and early nineteen-fifties was not without justification. Prior to the full flowering of molecular genetics spawned by Delbrück, Luria, Lederberg, and others, classical biochemistry was often impoverished by the dearth of correlative genetics required to lend confirmatory biological relevance to observations from the test tube. Stimulated by his encounter with Schrödinger's book, Haynes cultivated his growing fascination with biology. In particular he became interested in the physical forces involved in chromosome movement in metaphase and anaphase of the cell cycle, a problem in rheology that he considered to be more important than blood flow. In view of this newfound interest, Burton introduced Haynes to his friend and colleague, Raymond E. Zirkle at the University of Chicago, an authority on the mechanics of chromosome movement and a prominent radiation biologist. Zirkle had served as principal biologist for the Manhattan Project and he established a committee on biophysics (subsequently elevated to full departmental status) at the University of Chicago. This department included in its ranks the noted histologist William Bloom, as well as Peter Guiduschek, Robert Haselkorn, and David Freifelder, all of whom became prominent molecular biologists in their own right.

Following a brief fling in Europe as a postdoctoral fellow, Haynes was offered and accepted a junior faculty position with the Chicago group. In 1959, he set up shop in a laboratory recently vacated by Aaron Novick and adjacent to Robert Uretz's radiation microbiology facility. Together with Uretz, Haynes and his student Michael Patrick initiated studies on a dye-sensitized photodynamic inactivation system developed by Uretz and David Freifelder, using the yeast *Saccharomyces cerevisiae* as a test organism. Zirkle was a trained botanist and his familiarity with fungi likely motivated his choice of yeast as a test organism in his radiobiology research program. Uretz reasoned that if they perturbed yeast cells with acridine orange and other stains known to bind to nucleic acids, and then exposed the cells to light at wavelengths that were absorbed by the dyes, they might acquire a sensitive means of probing specific molecular components of the cell and of developing chemically defined ways of inactivating cells. In the course of these experiments, Haynes and Uretz in essence rediscovered the phenomenon of postirradiation recovery of cells, which by this time was graced in radiobiological circles with the formal designation of *liquid holding recovery*. Haynes recalled these events in his recent historical writings:

> Late one Friday afternoon, very early in our collaboration, Uretz and I completed a long and tedious series of X-irradiations of a suspension of stationary-phase diploid yeast. It was late in the day, and being tempted by the prospect of a visit to the Faculty Club bar, we decided to postpone plating the cells until the following Monday. After all, the cells were not growing and we felt sure they would not die, or be adversely changed, by storage over the week-end in our darkened laboratory. Accordingly, we left for the Club and plated the cells on rich growth medium when we returned Monday morning. A few days later when we took the plates out of the incubator, I got the shock of my life: they were covered with thousands of colonies instead of the 200 we were expecting [based on designed dilutions]. We could not believe we had made sufficiently serious dilution errors to get such a result. We immediately repeated the experiment, once again delaying plating a few days but using a series of plating dilutions in order to get an accurate estimate of the magnitude of the "recovery" effect. It emerged that it was surprisingly large.

Haynes was quick to point out that when he began this research his thinking and reading were not directed to radiation biology in general, nor to postirradiation recovery phenomena in particular. He told me that he "was for the most part unfamiliar with the large number of similar phenomena that had been reported previously in other organisms, dating back to the early observations of Hollaender in the nineteen-thirties." Indeed, as he subsequently discovered, liquid holding recovery in yeast after exposure to ionizing radiation had been independently discovered by the Russian radiobiologist V. I. Korogodin some years earlier. In recounting the history of these efforts in an article published in 1993 and entitled "The Study of Post-Irradiation Recovery of Yeast: The 'Premolecular Period,'" Korogodin, then at the Joint Institute for Nuclear research in Dubna, Russian Federation, traced the history of postirradiation recovery in yeast as far back as 1920 when G. A. Nadson first suggested that yeast cells can recover from so-called "radiation sickness." In his historical review, Korogodin reflected that "if one views molecular biology as a tool, such as a spade, then it is the study of phenomenology that shows where to dig." Korogodin also cited a paper published in 1949 by F. B. Sherman and H. B. Chase, who reported that "if yeast were X-irradiated and then kept in water, the number of viable cells increased." I shall shortly mention yet another important paper on this phenomenon published in that year.

Parenthetically, Haynes takes credit for rescuing Korogodin's contributions from the Russian literature. While browsing in the library one day at the University of Chicago, he encountered a copy of a Russian popular science magazine called *Paroda*, literally translated as "science." Since the magazine was entirely in Russian, he simply flipped through the pages. "But," he related to me, "my attention was riveted by a series of graphs that looked to me for all the world like a paper on yeast recovery." Haynes eagerly sought out a Russian/English dictionary, looked up a few key words, and excitedly concluded that this was indeed a paper on liquid holding recovery, very similar to the work that he and Patrick were then engaged in. "In our first paper on the topic, Mike Patrick and I made reference to six or eight of the Korogodin papers," he later told me.

Haynes only became aware of Hollaender's early work on post-UV recovery when he talked with him at the Annual Meeting of the Radiation Research Society in May 1962. At the time, Haynes and Patrick were experiencing considerable difficulty in reproducibly observing liquid holding recovery, a problem apparently shared by other investigators that certainly contributed to the general climate of skepticism about the biological significance of postirradiation recovery phenomena in general. When he confided to Hollaender that it was "very much a Monday, Wednesday, Friday type of affair," Hollaender told him to "check the water." And indeed the rigorous use of triple distilled water cured the reproducibility problems, at least in Haynes's experiments.

The significance of the liquid holding experiments conducted by Haynes and his colleagues lies not in the primacy of the observations themselves, which by his own admission were not at all novel, but in the particular interpretation that he placed on them. The interpretation of survival curves following the exposure of microorganisms to UV radiation had been undergoing important transition for some years. As Phil Hanawalt commented to me during one of our many conversations on this topic: "At the time that I and other graduate students joined the biophysics programs at places like Yale and Chicago in the early nineteen-fifties, many of the radiation biologists were still heavily enamored with classical target theory and interpreted survival curves, especially for the killing of mammalian cells, almost exclusively in this context." In its simplest form, target theory held that one could use radiation (and a hefty dose of mathematics) to calculate the size, and hence infer the nature, of the radiation-sensitive targets in cells. The smaller the target, the more radiation was required to kill cells. So, for example, the broad shoulder region of the survival curve that reflects relatively infrequent cell killing at low doses of radiation, and which typically precedes the more dramatic exponential inactivation region of the curve, was for a long time interpreted as an indicator of either the number of radiation "hits" required to inactivate the nucleus of the cell (so-called "multi-hit requirement") or the presence of multiple identical nuclei in cells, only one of which was necessary for survival (so-called "multi-target requirement"). The existence and operation of DNA repair processes was essentially ignored. Hanawalt told me:

> When one entered the Yale biophysics program, D. E. Lea's *Actions of Radiations on Living Cells* [which was published in 1946 and included a prominent exposition of target theory] was required reading. But an important piece of work that changed many people's views about this, including my own, was published by Richard Roberts and Elaine Aldous of the Carnegie Institution of Washington in 1949, the same year that Kelner's photoreactivation paper was published. In this work Roberts and Aldous held UV-irradiated *E. coli* cells in buffer for different periods of time before plating them and observed dramatic differences in the *shapes* of the survival curves. The fact that one could change the shape of a survival curve simply by

altering the time of holding the irradiated cells in buffer wiped out the validity of target theory for many of us.

Hanawalt was not alone in his skepticism about target theory. In typical unequivocal fashion, John Cairns ventured to me that:

> The hopelessness of the field of theoretical biology is nowhere better demonstrated than in the history of the target theory of radiation killing of cells. It was all absolute nonsense! But books and theses and papers poured out, and countless people were kept employed building castles in the air.

Even Walter Harm, who in a far more moderate vein suggested that "in many cases the principles of target theory can be applied in the quantitative evaluation of dose-effect curves (such as survival curves) provided one is fully aware of the modifying influence of repair processes," admitted that "target theory was...of limited use."

The paper by Richard Roberts and Elaine Aldous referred to by Hanawalt is indeed an important one in the history of radiation biology and possibly represents one of the more thoroughly documented studies on the phenomenon of liquid holding recovery. Their observations led Roberts and Aldous to be unambiguously explicit in their misgivings about interpreting survival curves based exclusively on the number of "hits" required to inactivate a radiation-sensitive target. Consider the following statement from their paper:

> Figure 7 shows survival curves obtained by plating after increasing intervals of the recovery treatment. The form of the curve shifts from concave to convex, passing through a straight line condition. It would obviously be inadvisable in this case to interpret the straight line condition as evidence for "single hits" in a homogeneous population. ...it is certain that the shape of the survival curve depends considerably on the exact experimental conditions and any interpretations based on the shape of the survival curves should be made with caution.

Hanawalt recalls that when he entered Yale as a graduate student in biophysics the "young Turks in the department, Dick Setlow, Walter Guild, Frank Hutchinson, and Harold Morowitz, were breaking free from this paradigm." But the real explanation for the dramatic shifts in the shape of survival curves after holding cells in liquid remained elusive, and according to Haynes "even as late as the early nineteen-sixties many radiation biologists still regarded the recovery phenomena exemplified in liquid holding experiments as poorly understood incidental consequences of radiation exposure." In fact, Haynes recounted that one of the reasons that Bob Uretz opted not to continue working on liquid holding recovery in yeast was that his view was that it was "unimportant." Haynes told me that Uretz thought that it was "just another one of those recovery processes of which there were a lot around." In a historical review published in 1993, Haynes wrote that:

> The discovery of liquid-holding recovery in yeast was the most important factor in stimulating my interest in studying DNA repair .... Most of the previous work on recovery and reactivation phenomena in irradiated cells could be interpreted only in terms of the formal concepts of target theory. In part, this was because no one knew specifically what the sensitive targets might be [for X rays], even though Zirkle's partial cell irradiation studies (among others) indicated that in eukaryotes, the nucleus was more sensitive than the cytoplasm. It had of course been suggested

that DNA was an important target, certainly for UV, and quite plausibly for X-rays. [But] when I was in England [as a postdoctoral fellow] there was much debate over the role of DNA as a target for ionizing radiation. However, I was one of those who supported the DNA hypothesis, and during the early 1960's more and more evidence seemed to be accumulating in favor of it. Thus, when we observed similar radiobiological responses in yeast and in *E. coli* for X-rays, as well as UV and nitrogen-mustard (for which the question of a DNA target was less controversial), we simply assumed that DNA damage was indeed involved in all these effects. Thus, we began to speculate about possible mechanisms of repair on the basis of our knowledge of DNA structure.

Haynes and Patrick reported their liquid holding recovery findings at the annual meeting of the Radiation Research Society in 1962. From the perspective of the history of the discovery of excision repair of DNA, the significance of this report lies in their attempt to explain the process by their allusions to a putative light-independent (dark) repair process. The authors entitled their abstract "The Possibility of Repair of Primary Radiation Damage in Yeast" and wrote that:

> If the plating of a starved suspension of diploid yeast (*Saccharomyces cerevisiae*) is delayed for some hours after irradiation in phosphate buffer or distilled water with either 150 kv X-rays or 2357 Å ultraviolet light, then the number of cells capable of forming visible colonies on wort agar is greatly increased. . . . All ploidies (haploid through hexaploid) show this recovery from UV inactivation; furthermore it is independent of and additive to photoreactivation. In all instances, the recovery process appears to act as a simple dose-modifying agent over several decades of survival . . . . Since recovery or rescue phenomena produced by procedures which alter metabolic rates after irradiation seldom act as true dose-modifying agents, it is not implausible that this slow recovery in buffer is based upon a direct physical enzymatic repair of the primary molecular lesions.

During our many discussions on the historiana of excision repair, Haynes stressed the significance of these experiments in the context of the "paradigm shift" (a phrase that he alluded to often) that unfolded in the period immediately following the elucidation of enzymatic photoreactivation as a DNA repair phenomenon. He related:

> Prior to 1957 there were a variety of phenomena out there, all of which were considered somewhat minor, but when taken together they provided the phenomenological background for thinking about other forms of DNA repair. Liquid holding recovery prompted us to consider that DNA is the target for damage and its repair. We knew about pyrimidine dimers at this time. So a group of us stood around a blackboard one day and wrote down three hypothetical mechanisms for repair of UV radiation and X-ray damage. Photoreactivation was obviously one of these, and we concluded that it was formally possible that some sort of X-ray damage might be directly reversed in the same way that dimers were. We entertained recombination as a second possible mechanism of repair. This was especially attractive to us because yeast cells are diploid and so we reasoned that recombination between sister chromosomes might mitigate against DNA damage. But we were also struck by the fact that DNA was a double-stranded complementary structure with informational redundancy. Given the complimentary base pairing in DNA and the informational redundancy we thought it possible that cells might actually be able to

cut out one damaged strand and use the opposite strand as an informational template.

In a paper published with his colleague W. R. Inch in 1963, in which Haynes elaborated on the reactivation phenomena already commented on by Harm, he wrote that:

> At least two general reactivation mechanisms may be involved: (a) functional elimination of the damaged region of the phage DNA by genetic exchange with homologous regions of the bacterial DNA; and/or (b) restoration of the damaged region by direct enzymatic repair or local DNA synthesis.

Haynes recalls that in the autumn of 1963 Dick Setlow visited the University of Chicago to give a seminar:

> I remember him coming into my office—we knew each other because we were members of the old radiobiology club and the biophysics club—and he triumphantly said to me, "Do you know how *E. coli* repairs UV damage?" I said, "No, how?" He replied, "It cuts out dimers and throws them away," or words to that effect. I realized that was a remarkable experimental observation, but it didn't really hit me as "Wow!" because we had thought of this idea independently and other people had thought of it too. But Dick was the first to get the important experimental data that proved this beyond a doubt!

Others had indeed thought about excision repair. In a paper published in the journal *Radiation Research* in 1963, Walter Guild, then in the Department of Biochemistry at Duke University, also speculated in general terms about radiation recovery mechanisms. Guild was familiar with a Ph.D. thesis by one J. H. Stuy, completed in 1961, which documented both the degradation of DNA in UV- or X-irradiated cells by a hypothetical nuclease activity, and the partial resynthesis of DNA "provided the dose is not too large." Guild wrote that:

> An entirely speculative mechanism based on these observations is that the presence of damaged DNA acts as a substrate-inducer to elicit an adaptive enzyme formation—namely a specific nuclease. This nuclease removes the damaged strands of DNA, along with considerable adjacent regions, preparatory to resynthesis of new DNA from the remaining intact strand.

In a letter he wrote to me in the spring of 1995, Haynes provided the following summation of the complicated historical events that in his view surrounded the elaboration of excision repair of DNA.

> It is possible that the concept of [excision repair], at least in rough outline, arose independently, and at roughly the same time, in the laboratories of Harm, Guild, Setlow and myself. What is interesting is that all of us knew one another and attended meetings of the Radiation Research Society and the Biophysical Society. Because this idea seemed to be so speculative in 1962/63, before Setlow's experiments, it is possible that none of us were sufficiently excited to mention it to one another. However, I do know that Dick [Setlow] was very excited by the result of [his] excision experiment and I became excited when he visited us at Chicago for a seminar in the autumn of 1963. It became immediately clear to me that he had critical biochemical evidence for this mechanism upon which the above-mentioned people merely speculated.

I first met Dick Setlow in 1966, an occasion of which I am sure he has little recollection, since it was during one of his many seminar visits around the country giving the latest on excision repair in *E. coli*, while I was but a budding postdoctoral fellow in David Goldthwait's laboratory in the Department of Biochemistry at Case Western Reserve University. As a physician trained in South Africa in the classical British mode, I was a recent convert to, and a novitiate of, the discipline of molecular biology. I had just made the transition from the starched white uniform of a pathology resident at Cleveland Metropolitan General Hospital to the trademark T-shirt and jeans of "hip" wet laboratory bench workers in the late nineteen-sixties. I was agog with my new work environment. Boris Ephrussi and his wife, Harriet Ephrussi-Taylor (household names in molecular genetics), worked in the building just across the street, and my ego was boosted by the fact that Larry Astrachan, then in the Department of Microbiology at Case Western Reserve, actually knew me by name, for no other reason than the fact that I had expressed some interest in doing postdoctoral training in his laboratory. But the genetics of the *rII* region of phage T4 proved too intimidating to me at that stage of my career and I selected Goldthwait's laboratory instead. I knew of Astrachan because he had worked with Elliot Volkin at Oak Ridge National Laboratory. The pair had achieved a measure of notoriety for having experimentally demonstrated the existence of mRNA in the mid nineteen-fifties, but failed to recognize this (as did the rest of the molecular biological community for quite a while). By way of an interesting digression, this story, fully recounted by Horace Judson in *The Eighth Day of Creation*, goes something like this.

In the summer of 1956 those in the vanguard of the "new" biology were deeply perplexed about how information was transmitted from DNA to proteins. Judson tells us that:

> The summer of 1956 was a good one for conferences, and communication was growing so intense—as had happened among the physicists of the twenties, too—that the conference net took on an importance it normally doesn't have. The Cold Spring Harbor Symposium, about genetic mechanisms that year, was merely a curtain raiser to a formidable week at the McCollum-Pratt Institute at Johns Hopkins University, in Baltimore, on "The Chemical Basis of Heredity," followed by the Gordon Conference on nucleic acids and then a shorter meeting, chiefly about proteins, at the University of Michigan in Ann Arbor: they merged all into one, a floating poker game with players dropping in or out.

This is the very meeting at which Sol Goodgal presented evidence for the existence of photoreactivating enzyme in *E. coli* (see Chapter 2). The final day of the meeting in Baltimore featured a last minute talk by Elliot Volkin. Volkin and Astrachan had been carrying out experiments at Oak Ridge in which they had pulse-labeled nucleic acids with ^{32}P immediately after infection of *E. coli* with phage T2, and then followed the fate of the label by chasing with cold phosphate. They separated labeled RNA from DNA and noted that some of the former species turned over very rapidly. This observation was not all that remarkable. Others had noted that a minor fraction of newly synthesized RNA turns over rapidly. What was remarkable and also very puzzling was their finding that the base composition of some of the newly synthesized RNA was exactly that of phage T2! According to Judson, Volkin and Astrachan were "apologetic to an extreme about the possibility of contaminated experiments

[and] uncertain about conclusions." They believed that they may have found an RNA precursor for DNA synthesis, an idea voiced a decade earlier by the Belgium biochemist Jean Brachet. Judson summed up the immediate reception to Volkin and Astrachan's findings as follows:

> The notion of an RNA precursor to DNA did not fit the biochemistry of DNA synthesis that Kornberg had [recently] elucidated. The Volkin and Astrachan RNA was an anomaly. It could not quite be dismissed as the inconsequential and self-admittedly sloppy work of people not well known at a lab not highly regarded.

A later publication by Volkin and Astrachan confirmed the results with cleaner experiments. Still, they (as well as the upper echelon of molecular biology) failed to see the light. In fact it was not until the existence of mRNA was firmly established that these experiments were finally correctly interpreted.

Of course Astrachan knew Dick Setlow from his days at Oak Ridge. He therefore invited Setlow to talk at Case Western Reserve in 1968 and asked several postdoctoral fellows and graduate students to meet with Setlow at his home the evening after Setlow's seminar. I remember very little about that social occasion except the extension of my sense of general excitement, and my growing conviction of the merits of switching from the dreariness of the pathology morgue to the lively excitement of the research bench. I also recall very little about the specifics of Dick's seminar. I certainly did not make a conscious decision at that time to become a DNA repair "groupie." But, in retrospect, the amalgamated experience of Dick Setlow's seminar and meeting him in person must have imprinted itself on my subconscious in more than a casual way. Following my postdoctoral training with Dave Goldthwait, I (and as recounted in the next chapter, several others) went in pursuit of the repair enzymes that participate in the biochemical pathway identified by Setlow's investigations.

Dick Setlow began his professional career as a graduate student in the Department of Physics at Yale. When I asked him whether he too was seduced to biology by Erwin Schrödinger's little book, he refreshingly replied, "No, I thought that physics was sterile in the early nineteen-forties. People were doing the same things over and over again and I was looking for a change." An opportunity for change presented itself in the form of the biophysics program at Yale that was being established at that time by Ernest Pollard. Some years later, when Phil Hanawalt was seeking advice about graduate programs in biophysics, a physicist from the University of Michigan, Robley C. Williams, recommended the biophysics program at Yale to him, an event that I will shortly recount in greater detail. Among other achievements, Pollard, a nuclear physicist, had built a cyclotron at Yale with the intention of investigating the effects of charged particles on biological systems. This cyclotron has an amusing history that was recounted to me by Hanawalt. When the famous American nuclear physicist Ernest O. Lawrence was at Yale he requested funding from the University to construct the first cyclotron, but was denied. This (perhaps abetted by other influences) prompted him to leave Yale for the University of California at Berkeley which considered Lawrence's request favorably. In fact, Yale did not acquire a cyclotron until after Lawrence obtained a second larger machine at Berkeley and he gave Pollard the inner chamber from his first one, which he (Pollard) used to construct Yale's first such machine at a cost not much less than that originally estimated for a brand new one! "So Yale eventually acquired this little cyclotron that was no more energetic than Lawrence's first machine," Hanawalt recalled, "and it really wasn't of much value to many nuclear physicists, who preferred to go to Berkeley to use the big one."

While in graduate school during World War II, Setlow was also an instructor in the physics department at Yale, and upon completing his graduate studies, he joined the faculty of the department, where he taught atomic and nuclear physics, biophysics, and eventually a graduate course in spectroscopy. "This," he told me, "was my specialty. I was a classical physical spectroscopist." In the early nineteen-fifties, Setlow became interested in examining the effects of monochromatic UV radiation on biological macromolecules. Together with Donald Fluke, a physics graduate student with an interest in biophysics, he constructed several large water prism monochromators. "They were the biggest monochromators going at the time," he told me. As noted in an earlier chapter, the ability to correlate the action spectrum of a given biological effect with the absorption spectrum of a particular molecular species was a recognized means of attributing particular functions to specific molecules. Setlow investigated the action and absorption spectra of proteins and found that in general they did not correlate well, because different amino acids have different sensitivities to UV radiation. This led him to shift his interest from proteins to DNA as a subject for his monochromators in order to determine whether the action and absorption spectra for DNA were similar. They were.

Setlow's interest in DNA damage and repair did not really materialize until he went to Oak Ridge National Laboratory for an intended one-year sabbatical, which evolved to a permanent move. "I was spending too much time teaching at Yale," he told me, "and I became more interested in doing research. They had a lot of good researchers at Oak Ridge. John Jagger was there and he and Alex Hollaender had set up a big monochromator." But Setlow considers that his "lucky" break in DNA repair was really his early introduction to thymine dimers:

> One of the people on the staff was David Krieg, a geneticist. He approached me one day and asked what I thought about the experiments carried out by "those crazy Dutchmen?" Of course he was referring to the experiments of Beukers and his colleagues showing the formation of thymine dimers when frozen solutions of thymine were exposed to UV radiation [see Chapter 2]. That was the first indication I had of the possibility of forming thymine dimers in ice. This observation motivated me to examine the wavelength dependencies of making, and primarily breaking, thymine dimers. Short wavelengths broke them and one could measure the action spectrum for that. As soon as I observed what the kinetics looked like it seemed to me that one should be able to make dimers in biologically active DNA with short wavelengths of UV radiation, thereby functionally inactivating the DNA, and then break the dimers with even shorter wavelengths thereby reactivating it.

Setlow and his colleagues, including his then-wife, Jane Setlow, tested this idea experimentally and happily obtained the proposed result. "We did more of these kinds of experiments," he recalled, "and showed that this type of dimer reversal overlapped photoreactivation. I became more and more convinced that dimers were biologically important." Setlow was a good friend of Ruth Hill and was fully aware that she had isolated a UV-radiation-sensitive strain of *E. coli* B. Additionally, as recounted above, Evelyn Witkin had established that *E. coli* B/r was the more appropriate wild-type control strain for UV radiation survival studies. So a perfect combination of strains were available to ask what, if anything, thymine dimers had to do with the inhibition of DNA replication that was known to occur in UV-irradiated cells. Immediately prior to Setlow's move to Oak Ridge, several laboratories were investigating the effects of UV light on the synthesis of macromolecules such as RNA and DNA. Phil Hanawalt, then a graduate student in Setlow's laboratory at Yale, obtained a novel mutant of *E. coli* from Seymour Cohen which had an auxotrophic requirement for thymine. This allowed him to label the DNA of this strain with radioactive thymine (by growing cells in the presence of this compound), thereby providing a highly sensitive and convenient method for quantitating DNA synthesis. During these and other studies that Hanawalt and Setlow reported in 1960, they confirmed earlier observations that when *E. coli* cells were exposed to UV radiation they underwent rapid inhibition and subsequent recovery of DNA synthesis.

In the early nineteen-sixties Setlow, together with his colleagues Paul Swenson and Bill Carrier, examined the effects of UV radiation on DNA synthesis in *E. coli* B/r and B_{s-1} and simultaneously monitored the fate of pyrimidine dimers in these two strains. Using the now well tried method of incorporating radioactive thymine to measure the kinetics of DNA synthesis, they observed that DNA replication in the UV-sensitive B_{s-1} strain was inhibited by minute amounts of UV radiation compared to the UV-resistant B/r strain. More importantly, they noted that whereas in the latter strain DNA synthesis recovered its normal rate after some period of time, *E. coli* B_{s-1} cells did not. Setlow was convinced from his previous studies on the formation and breakage of dimers by different wavelengths of UV light and from his own studies on enzymatic photoreactivation that the inhibition of DNA synthesis was the result of the blockage of DNA replication by thymine dimers. Indeed, at about this time he and his colleague at Oak Ridge, Fred Bollum, showed that inhibition of the synthesis of UV-irradiated DNA could be reproduced in vitro using a purified UV-irradiated DNA template and a DNA polymerase that Bollum had purified from calf thymus. Setlow, Swenson, and Carrier therefore suggested that dimers blocked DNA synthesis in both *E. coli* B/r and B_{s-1} cells. But whereas the normal strain was able to recover from this effect, the UV-sensitive mutant was not.

The Setlow group showed that the inability of B_{s-1} cells to recover from the effects of UV irradiation was not due to a defect in photoreactivation. When cells were exposed to photoreactivating light immediately after UV irradiation, the inhibition of DNA synthesis was mitigated in both the B/r and B_{s-1} strains to an equal extent. Firmly persuaded that the recovery of DNA synthesis reflected some other sort of repair of pyrimidine dimers, they stated in a paper published in 1963 that:

> Since thymine dimers are not split in the resistant cells, we are investigating other molecular mechanisms that may account for the resumption of synthesis, such as the possibility of polymerizing at a slow rate around a block, *or that the dimer is cut out of the DNA chain by nucleases and is replaced by two thymines* [my italics].

Setlow's conviction about the existence of other mechanisms for the repair of thymine dimers stemmed in part from the prevailing intellectual climate on

DNA repair discussed earlier. This is reflected in the continuation of the quotation from the 1963 *Science* paper just provided:

> The first possibility seems to exist in a system which is in vitro. The second is consistent with the postulated mechanisms concerning host-cell reactivation and ultraviolet reactivation of bacteriophage assayed on irradiated and unirradiated sensitive and resistant mutants. [A reference to phage work of Walter Harm published in 1963 is provided here.]

Similarly, in the opening sentence of their classic paper published in 1964 entitled "The Disappearance of Thymine Dimers From DNA: An Error-Correcting Mechanism," which marked the first formal experimental demonstration of the excision repair of pyrimidine dimers, Setlow and Carrier wrote that: "Recovery processes associated with ultraviolet irradiation (e.g., photoreactivation, heat reactivation, photoprotection, liquid-holding recovery, host-cell reactivation, and UV reactivation) probably act enzymatically."

In our several discussions, Setlow readily acknowledged that "all these other recovery and reactivation phenomena clearly existed at that time." But phenomenology is one thing and definitive experimental evidence for a new DNA repair mode was something else entirely. When, during the course of their experiments on DNA synthesis with *E. coli* B/r and B_{s-1}, Setlow and his colleagues measured the actual content of thymine dimers in whole cells they found to their disappointment that dimers persisted in the UV-resistant B/r cells during incubation in the dark, despite the fact that the cells had undergone recovery, as evidenced by their capacity to carry out DNA synthesis. This presented a troubling paradox. If dimers really were responsible for the biological effects of UV radiation, and the radiation-resistant B/r cells were less sensitive to these effects by dint of some sort of specific repair mechanism, then B/r cells must surely have a way of getting rid of dimers from their DNA. Yet measurements of the thymine dimer content in cells as a function of the postirradiation incubation time indicated exactly the opposite. Setlow solved this paradox in a leap of intuition based on his interpretation of a key experiment. He recounted this important intuitive leap to me as follows:

> We did an experiment in about April 1963 in which we UV-irradiated *E. coli* B/r and let the cells incubate for some time in the dark [i.e., in the absence of photoreactivating conditions]. At the end of that time we exposed the cells to photoreactivating conditions and then harvested them and measured their thymine dimer content. All the dimers were still in the cells—they had not been reversed. Whereas when we did the same experiment with the B_{s-1} strain all the dimers were reversed by photoreactivation. This result suggested to us that in the B/r strain the dimers were in a state in which for some reason they were not accessible to the endogenous photoreactivating enzyme. I distinctly remember telling a scientific review committee that was visiting Oak Ridge at that time that we didn't know exactly what was going on, but that I was convinced that there was something different about the state of the thymine dimers in the two strains because of the lack of photoreactivability in *E. coli* B/r. The other relevant thing that we knew then was that photoreactivating enzyme from yeast didn't work as well on single-stranded DNA as it did on double-stranded DNA, but we didn't immediately relate this to our observation in *E. coli* B/r. We also knew a fair amount about the action of various types of nucleases on DNA containing thymine dimers from experiments that we had carried out in collaboration with Fred Bollum using oligonucleotide polymers that he used for his DNA polymerase experiments. We knew for example

that the limit digest of an oligonucleotide containing dimers by snake venom phosphodiesterase was a trinucleotide containing a dimer. So we reasoned that if thymine dimers were cut out of the DNA during some sort of repair mode they would likely be in the form of small oligonucleotides, not as free thymine.

All of these tantalizing bits of information provided the nub of an idea about how excision repair might transpire in UV-irradiated *E. coli*, leading Setlow to the notion that perhaps dimers were insensitive to photoreactivation in *E. coli* B/r because they were cut out as part of a single-stranded oligonucleotide fragment. This in turn led to the conclusion that instead of measuring the thymine dimer content of *whole* cells it might be more informative to fractionate the DNA into its acid-insoluble and acid-soluble components, which allows the separation of small oligonucleotides (acid-soluble) from the bulk of the acid-insoluble DNA. He credits some of this experimental reasoning as yet another bonus of his collaborations with Fred Bollum. Setlow related:

> Fred did this sort of fractionation of acid-soluble and acid-insoluble DNA all the time in his experiments with DNA polymerases. After he'd tested a DNA polymerase in vitro in some sort of DNA replication experiment, he'd put the entire reaction mixture in acid and examine what had been incorporated into the acid-insoluble fraction of DNA. It's very easy to do.

It is not difficult to understand how all these contemporaneous observations eventually led Setlow to the idea that perhaps what was different about the response of thymine dimers to photoreactivating conditions in the UV-resistant B/r cells was that during the preceding incubation in the dark they were excised from the DNA as components of small single-stranded oligonucleotides. When, instead of measuring the thymine dimer content of whole cells, Setlow and Carrier broke the cells open by sonication, thereby releasing their contents, they observed that the thymine dimer content of the fraction that was precipitated by trichloroacetic acid (the so-called TCA-insoluble fraction, which includes all the high-molecular-weight DNA) was indeed reduced in B/r but not in B_{s-1} cells. Furthermore, as an incisive experimental corollary of this observation, they noted that in the former cells the dimers could be recovered in the TCA-soluble fraction containing small oligonucleotides, conclusively indicating that they had been physically excised from the DNA. The latter experimental result was crucial to the formal demonstration of excision repair, since the unadorned observation that thymine dimers disappeared from acid-insoluble DNA when UV-irradiated cells were incubated under nonphotoreactivating conditions was clearly open to other mechanistic interpretations about their repair.

The technique then in use for measuring the thymine dimer content of acid-precipitated UV-irradiated DNA was complicated, time-consuming, and also quite dangerous. I learned it firsthand from Betsy Sutherland, one of Dick Setlow's former graduate students, when Betsy (together with her husband John Sutherland) and I were at the Walter Reed Army Institute of Research, an unwelcome twist in my career that I shall briefly recount later. After harvesting the precipitated DNA, it was necessary to reduce the radiolabeled thymine dimers and monomers to their free state so that they could be separated and quantitated by two-dimensional paper chromatography. This required the complete chemical hydrolysis of the precipitated DNA in formic acid at a temperature of about 176°C. Formic acid/DNA solutions representing each data point were transferred to thick-walled glass tubes, which were sealed by flaming them and carefully drawing out the melted tops. If one carried out this sealing operation correctly, the tubes survived the very high temperature perfectly well. However, if one was inexperienced or inept in this tricky maneuver, the tubes often exploded in the hydrolysis oven, or even worse, very soon after their removal from the oven. Opening sealed hydrolysates after removing them from the oven was an event that I (and my students and technicians) anticipated with considerable trepidation, despite the protection afforded by plastic face masks, goggles, and thick gloves. Nothing cleared the laboratory faster than the announcement that someone was about to open boiling hot hydrolysis tubes! Most students of excision repair were enormously relieved when in later years alternative techniques for measuring thymine dimers emerged.

Based on their hunch that if thymine dimers were indeed released into the acid-soluble fraction of DNA they would likely be in the form of small oligonucleotides rather than free thymine, Setlow and Carrier realized that the oligonucleotide state would preclude the measurement of thymine dimers by the standard chromatographic separation procedure they employed. So they subjected this fraction to hydrolysis at high temperature as well in order to render free thymine and thymine dimers. This technical caveat turned out to be most significant as we shall see when we examine the contemporaneous studies of Richard Boyce and Paul Howard-Flanders at Yale. Setlow and Carrier also enjoyed a measure of sheer good fortune that is often an essential ingredient of success, but which one typically only recognizes in retrospect. The maximal size of oligonucleotides that permits their solubility in TCA is about 10 nucleotides. Many years after the formal discovery of excision repair, it was shown that this is just about the size of the fragments in which dimers are excised in *E. coli*. Had the excision fragments been much bigger than this they may have precipitated with the rest of the DNA in the TCA-insoluble fraction and the formal discovery of excision repair would have required some other experimental approach.

In their 1964 paper in which they announced excision repair in *E. coli*, Setlow and his collaborators stated that: "Although these data indicate nothing about the mechanism of dimer removal, nor what, if anything, takes their place in the DNA, it is reasonable to suppose that DNA synthesis resumes in resistant cells because thymine dimers are removed."

It would take another 20 years for the precise mechanism of dimer removal to be fully elaborated, a history that will not be recounted here. But as early as 1965, when summarizing a symposium on *Structural Defects in DNA and Their Repair in Microorganisms*, Setlow prophetically volunteered that the disappearance of thymine-containing dimers from the acid-insoluble fraction and their appearance in the acid-soluble fraction "represents two events—a right-hand cut and a left-hand cut." This turned out to be exactly the case. Setlow and Carrier also speculated about the generality of excision repair in their 1964 paper: "In addition, the processes we have observed might be typical of all error-correcting mechanisms involving DNA chains of unnatural or non-Watson-Crick structure." Finally, a new and long-anticipated DNA repair pathway was uncovered!

The literature on excision repair of DNA appropriately attributes contributions to this discovery by Paul Howard-Flanders and his postdoctoral fellow Richard P. Boyce, and by Philip Hanawalt and his student David Pettijohn. What is perhaps not as well recognized is that both Hanawalt and Boyce had strong historical ties to Setlow as he was their former mentor, and hence they were inclined to seek him out for advice and consultation. Furthermore, on the basis of his own results, Setlow was in a unique position to provide crucial interpretative insights to the independent experimental observations made by both of these groups, as indeed he did.

My long-standing professional and personal association with Phil Hanawalt has been motivated by our mutual interest in DNA repair, our love of teaching, and our passion for writing, shared and embellished during the 19 years between 1971 and 1990 that we served together on the faculty at Stanford University. I first met Phil in 1969 when he visited the Division of Nuclear Medicine at the Walter Reed Army Institute of Research in Washington D.C. The previous year I had completed my postdoctoral fellowship training with Dave Goldthwait at Case Western Reserve University and was eagerly anticipating launching my career as a member of the faculty in the Department of Pathology at Stanford University, where David Korn, the new chair of pathology, was attempting to build a research environment with a solid foundation in molecular pathology. I suppose that David viewed someone who had formal training in classical human pathology and was also working on the "molecular pathology" of phage and bacteria as appropriate to such a mission. However, literally weeks prior to my planned departure from Case Western Reserve University for California toward the end of 1968, I received the astounding news that I had been drafted into the U.S. Army Medical Corps as a pathologist. This news was doubly shocking to me because prior to emigrating to the United States from South Africa in 1965, I was informed that I was beyond the age of eligibility for military service. But the escalating American involvement in Vietnam changed all that in one fell swoop of Congressional legislation, and like thousands of other physicians operating under the sublime assumption that they were "safe" from the draft, I soon found myself in a captain's uniform moodily gracing the classrooms and training fields of the U.S. Army Medical Corps at Fort Sam Houston, Texas.

Not being a U.S. citizen at the time that I received a draft notice, I had the legal option of refusing this call to duty and of leaving the country (forever) without the stigma of formal draft evasion. Through a series of events, the details of which are now shrouded in nothing more identifiable than a general sense of panic bordering on hysteria, I journeyed to Washington, driving through the night. I was granted an interview with a senior officer in the Army Medical Corps who listened to my fervent plea to remain in a research environment and secured a commission for me in the Radiobiology Division of the Walter Reed Army Institute of Research in Washington, provided that I agreed to enlist for three years instead of the obligatory two. The choice was not a difficult one. Faced with the prospect of diagnosing malaria and other exotic tropical diseases in the steamy jungles of South East Asia, the decision to carry out research in some, indeed, any aspect of radiation biology in Washington before embarking to Stanford (where my faculty position was being held) was not a difficult one.

Over lunch with Hanawalt during a visit to Walter Reed at which he presented a seminar, I informed him of the recent discovery that I and my technician (a "lucky" drafted graduate student named Jack King) had made in our tiny makeshift laboratory. Having survived six weeks of basic training at Fort Sam Houston without acute alcoholic cirrhosis of the liver, I had decided to go in search of the enzymes that Dick Setlow's discovery of excision repair predicted. I made crude extracts of excision repair-proficient and deficient *E. coli*, incubated them in the presence of UV-irradiated radiolabeled DNA, and searched for evidence of preferential nicking of the DNA by the former extracts using alkaline sucrose gradient sedimentation. I reasoned that if I could detect such an activity I would have a reliable, albeit tedious, assay for purifying such an enzyme. Many dozens of sucrose gradients later, I had no hint of a UV-radiation-specific nicking activity in extracts of *E. coli*.

Since each sucrose run in the ultracentrifuge took at least three hours, Jack King and I had lots of time to watch the 1969 World Series, and on occasion, to read the literature on DNA repair. It was then that I encountered Walter Harm's observations on the relative UV resistance of phage T4 detailed earlier in this chapter. Additionally, I encountered a brief report among the published abstracts of the 10th Annual Meeting of the Biophysical Society held in 1966, written by Dick Setlow and Bill Carrier, that attracted my attention. Fully aware of Harm's pioneering studies on the relative UV resistance of phage T4 and its dependence on a single locus in the phage genome, Carrier and Setlow reported to the Biophysical Society their observation that *E. coli* cells infected with this phage readily lost dimers from the acid-soluble fraction of their DNA, whereas cells infected with the phage mutant isolated by Harm that had lost UV resistance (phage $T4_{v1}$) did not.

The first time that Jack and I carried out our standard experiment with extracts of *E. coli* that had been infected with phage T4 we observed dramatic nicking of the UV-irradiated DNA. No such nicking was observed with extracts of phage T2-infected cells, nor with extracts of cells infected with Harm's UV-sensitive mutant phage $T4_{v1}$. Here was the biochemical explanation for the enhanced UV resistance of phage T4 relative to other phages in the T series first observed by Salvador Luria almost 25 years earlier, and elaborated by Walter Harm. I recall my excitement in breathlessly confiding this result to Harm by telephone. "I know why phage T4 is more resistant than T2 to UV light," I told him. "The phage encodes an enzyme that specifically recognizes pyrimidine dimers and incises the DNA at these sites." Harm was pleased, but expressed little surprise!

It would take another 11 years to elucidate that the enzyme that King and I had discovered (which, as recounted in more detail in the next chapter, was independently discovered by Mutsuo Sekiguchi and his colleagues in Japan at about the same time) was not the same beast as that involved in the excision of pyrimidine dimers in *E. coli*. By some quirk of evolutionary selection, a gene in phage T4 and a similar gene in a radioresistant bacterium called *Micrococcus luteus* are the only two documented examples of genes that encode proteins which specifically and uniquely participate in the repair of pyrimidine dimers in DNA by a novel excision pathway. The biochemical pathway by which dimers are repaired by these two enzymes is quite distinct from that which operates in *E. coli*. I shall return to this aspect of the history of excision repair in the next chapter.

Hanawalt had more than a passing interest in our experimental results. He immediately recognized that the absolute specificity of the T4 enzyme for pyrimidine dimers made it a potentially highly sensitive probe for the presence of very small numbers of these lesions in DNA, a notion that he and his colleague Ann Ganesan at Stanford subsequently exploited in many interesting and informative ways, culminating in the elaboration of an elegant assay for the detection of dimers in different regions of the genome and in different strands of a single gene. This lunch at Walter Reed with its enthusiastic overtones remains for me a fond memory of the beginning of my long and productive relationship with Phil Hanawalt.

Immediately following his high school graduation in 1949, Hanawalt enrolled for a year in a little known undergraduate college cum ranch called Deep Springs near the town of Bishop, California. Hanawalt told me:

> I received an informational brochure on Deep Springs as a consequence of having received an honorable mention in the Westinghouse National Science contest for my experiments on Wheatstone bridge circuits. The place sounded novel. Everyone who was admitted got a full scholarship. One worked 20–30 hours a week on the ranch and took courses for the rest of the time. Robley Williams, a biophysicist and electron microscopist who was on sabbatical at Deep Springs from the University of Michigan, taught calculus there and also gave lectures in biophysics during the first semester. My first ranch responsibility was that of meat cutter and boarding house manager. Late one Saturday night I cut my finger quite deeply while slicing ham for Sunday breakfast. Robley Williams volunteered to drive me into Bishop, some 36 miles away, to have this attended to, and during the two-hour trip we talked about one thing and another. He told me that if I wanted to go into a really exciting area I should consider biophysics because the field was entering a stage of research in which the basis of life itself was beginning to be understood at a molecular level.

After completing this year at Deep Springs, Hanawalt transferred to Oberlin College where he completed his undergraduate education with a major in physics. In his one (and only) biology course at Oberlin in 1953, he learned that "DNA was one of the chemicals in chromosomes. That's how much they knew about DNA at Oberlin College in the year of Watson and Crick," he reflected sardonically. Hanawalt contacted Williams again, who reinforced his earlier advice to pursue biophysics and suggested Yale as a leading graduate school in that arena. An interview with Ernie Pollard convinced Hanawalt that Yale was indeed a fine place to study at the interface between physics and biology. At Yale, Hanawalt first came under the influence of Harold Morowitz, a young assistant professor who created yet another convert to Schrödinger's *What Is Life?* In Morowitz's laboratory, Hanawalt had his first introduction to the intricacies of synchronous growth in *E. coli*, an area of investigative activity to which he has returned periodically during his career. He recounted to me that:

> The first time I attempted to synchronize the growth of *E. coli* by glucose starvation, it worked perfectly. The reason was that, unknown to me, the can of pipettes I was using turned out to have a mixture of 0.1- and 0.2-ml pipettes. By randomly selecting pipettes, I inadvertently generated a growth curve that looked for all the world like almost perfect synchronous growth for two division cycles. But of course I could never repeat this.

One day, Pollard, then the chair of the Biophysics Department at Yale, summarily informed Hanawalt that Morowitz had too many graduate students and that he would have to transfer to either Setlow's or his (Pollard's) laboratory. Hanawalt elected Setlow as his mentor, having made his acquaintance during the excellent spectroscopy course that Setlow taught. In Setlow's laboratory, Hanawalt evolved a Ph.D. thesis project dedicated to exploring the effects of UV light on macromolecular syntheses, including the synthesis of DNA, RNA, proteins, and lipids. As already indicated, during the course of these studies he noted that DNA synthesis was inhibited roughly in proportion to the dose of UV radiation delivered and then recovered quickly, and that this inhibition could be substantially mitigated by prior photoreactivation. "We had no idea what the lesions were," Hanawalt confided. "So when we published this work we talked about recovery processes—we didn't use the word 'repair.' In fact, as I recall, I never used the term 'repair' while I was at Yale."

Prompted by his enduring primary interest in synchronous growth of *E. coli*, Hanawalt elected to carry out postdoctoral training in Copenhagen with Ole Maaløe, a world expert in this area. Upon arriving at the spanking new Microbiology Institute where Maaløe had just been made professor, Hanawalt was not directed to a specific research problem, so he decided to explore the problem of thymineless death, a problem that he had first encountered in Setlow's laboratory when he worked with the thymineless auxotroph of *E. coli* (obtained from Seymour Cohen) for radiolabeling DNA. Hanawalt recalled:

> Ironically, my analysis of thymineless death converged on the subject of synchronous growth through my discovery that the inhibition of protein synthesis allows completion of the replication cycle, but not the initiation of new rounds of DNA replication. But the most important thing about my postdoctoral experience with Maaløe was that I learned how to manipulate and handle bacteria and even occasionally looked at them under the microscope. I also improved my ping-pong game considerably because Maaløe was an avid player. He would take his postdocs up to a room on the top floor of the new research building we worked in. This was euphemistically labeled as a storage room but it housed nothing other than a full-sized ping-pong table. There Maaløe would systematically and methodically whip the pants off one while casually smoking an enormous cigar. Part of his winning strategy related to those damn cigars. The ash would get longer and longer but never seemed to drop. I would become increasingly fascinated with the prodigious length of the ash and eventually would lose my concentration on the ping-pong game!

Hanawalt wanted to have a second postdoctoral experience in the United States, where he reasoned that it might be easier to negotiate a faculty position. Maaløe organized a small but high-powered molecular biology meeting in Copenhagen while Hanawalt was still around. Among other scientific luminaries of the time, the meeting included Jim Watson, Max Delbrück, Andre Lwoff, François Jacob, Jacques Monod, Francis Crick, Matthew Meselson, David Phillips, John Kendrew, and Max Perutz. Hanawalt was suitably intimidated. "I was extremely shy and sat at the very edge of it all," he recalled. "However, I summoned up some of my courage when I found myself standing next to Jim Watson in the toilet, and so I asked him where he thought I should go for a second postdoc experience in the United States. He curtly replied, 'Heppel or Sinsheimer,' zipped up his fly, and left. This was the extent of my first conversation with Jim Watson."

Hanawalt completed a further year of training with Robert Sinsheimer at Caltech, where he became proficient in density gradient ultracentrifugation. Here he encountered Matthew Meselson and rekindled his interest in the effects of UV light on DNA replication. He was also encouraged by the fact that Delbrück was very interested in the discovery of thymine dimers by the Dutch workers and had established research groups in his laboratories at Caltech and later (as mentioned in an earlier chapter) in Cologne, Germany, to explore various aspects of photoreactivation and other UV reactivation phenomena. Hanawalt recalls that "Delbrück was so excited by the 'new photobiology' that he gave a course in it. The front row in his class included Bob Sinsheimer, George Hammond and the noted radiobiologist Harold Johns, to help keep him honest." Like others who encountered Delbrück in the course of their scientific careers, Hanawalt has his share of indelible memories about the man. In his own recent historical reflections he wrote that:

> Delbrück was one of the most intellectually honest individuals it has been my pleasure to know, a true scientist who could never accept a partial proof or conjecture as adequate. It was his impatience with mediocrity that led to his wide-spread reputation for "destroying" seminar speakers on the stage; if Delbrück did not understand something in the lecture he would interrupt and ask a question of the speaker. He could accept an "I don't understand either" in response but was incensed by any attempt to explain without adequate proof. In the latter case, the speaker would gradually and painfully dig himself deeper and deeper into a pit of inaccuracies and uncertainties in response to Delbrück's persistent attack.

Hanawalt was recruited to the faculty at Stanford University in 1961 by the late renowned radiotherapist and radiation biologist Henry S. Kaplan. Kaplan, then Chairman of the Department of Radiology at Stanford, had with his far-reaching scope and interest, established a Laboratory of Biophysics at Stanford, where Hanawalt initiated his research program. Initially motivated by his desire to apply the technique of density labeling to examine DNA replication, in particular the physical properties of the replication fork, Hanawalt found himself coming back full circle to the effects of UV radiation on this process, the topic of his Ph.D. thesis with Dick Setlow. In his recent historical retrospective he wrote that:

> I set out to use the technique of density labeling with the thymine analog 5-bromouracil to study DNA replication intermediates in *E. coli*. I wondered why such intermediates were not readily seen in CsCl equilibrium density gradients at positions halfway between the parental (thymine-containing) DNA and the hybrid band in which a 5-bromouracil daughter strand was paired with a thymine-containing parental strand. We eventually appreciated that this was in part because of the fragility of the DNA, particularly at replication forks. . . .Realizing that thymine dimers were blocks to replication I wondered whether we might catch a segment of DNA in the act of replication (i.e., partially replicated and therefore partially density-labeled) if that segment contained one or more pyrimidine dimers. David Pettijohn joined me as a graduate student, and we soon had obtained density-label profiles of DNA from UV-irradiated *E. coli* that had incorporated tritium-tagged 5-bromouracil. To our surprise, we found. . .[that] many of the newly replicated DNA molecules contained too little 5-bromouracil to shift their densities appreciably. The significance of this observation became evident just a few months later.

What transpired a few months later was a telephone call to his former mentor Dick Setlow to inform him of these results. Having just deciphered the excision repair of thymine dimers in his own laboratory, Setlow immediately recognized the significance of Hanawalt and Pettijohn's observations and dispatched the following note to his former graduate student on August 23, 1963:

> Dear Phil:
>
> Enclosed is a preprint of some of the work we have done on the effects of UV on DNA synthesis in bacteria. Some later results that I am just beginning to write up for publication fit very well with those you have obtained on bromouracil incorporation into bacterial DNA. These results indicate that thymine dimers in radiation-resistant cells are cut out of the DNA and appear in the acid-soluble fraction of the cells, whereas in sensitive strains they are not cut out. An obvious mechanism is that the lesion is removed and perhaps the surrounding polynucleotide regions are replaced by new bases from the medium. In this sense they act as if there had been a turnover in DNA—the turnover being initiated by nucleases acting on the DNA lesion. If this is true, and our data seem to indicate this, then the bromouracil would be distributed randomly along a single strand and one wouldn't expect to find much melting of the heavy label.
>
> Best regards,
>
> Richard B. Setlow

Hanawalt candidly confirmed to me that:

> I had no particular reason to think of dimers being cut out of DNA. I assumed that they were blocking replication and killing cells. So my first thought that this might be a repair process was when Dick suggested it. In fact, when I got this note from Dick I called him and asked whether he was sure that he was not simply observing DNA degradation, and he replied that he was certain that dimers were being selectively removed. As soon as we began to think that we might be observing repair synthesis, we did various control experiments and showed that this synthesis was reduced by photoreactivation and that it did not constitute any sort of end-addition of nucleotides to sheared fragments.

Hanawalt and Pettijohn published a preliminary report of their findings in 1963 stating that their results are consistent with the hypothesis that "some type(s) of photochemical damage to DNA may result in the partial replication of the DNA molecule." In a definitive exposition published a year later they related this nonconservative mode of DNA replication to excision repair:

> Our results indicate that the early DNA synthesis following u.v. irradiation involves short, single-stranded segments distributed at random in the u.v. damaged bacterial genome. The results of thermal denaturation suggest that the phosphodiester backbone of the molecules involved in this synthesis is intact. These findings, when combined with the demonstration that thymine dimers are excised from u.v. irradiated DNA... is strong evidence for the postulated "repair replication" of damaged DNA....Repair in this manner could account for biological phenomena such as host cell reactivation...and u.v. reactivation...dark reactivation...mutation frequency decline...and post-irradiation treatment effects on mutation frequency....

The third research group to score a bull's-eye in the target of excision repair in *E. coli* was that of the late Paul Howard-Flanders. Howard-Flanders was a prominent radiobiologist at Yale who, as we shall shortly see, independently evolved an experimental path leading to excision repair which intersected with that of Setlow and his colleagues. At about the same time, he also discovered a different cellular response to UV radiation damage that came to be known as *postreplication repair* or *recombinational repair*, an important historical nuance that in the interests of space I leave to future historians of the topic of recombination. Regrettably, Paul passed away in 1991, and although I knew him well in his later years, I never had the opportunity to discuss these events with him firsthand. However, much of the hands-on work in these research endeavors was contributed by his postdoctoral collaborator Richard Boyce, who provided me with a detailed and fascinating written account of his involvement in the excision repair story. Dick Boyce thought well of Howard-Flanders and had great respect for his intellect as a scientist. Boyce told me:

> Paul's Quaker upbringing was reflected in his personality. He was the kindest and most compassionate man I have known. He was a pacifist and absolutely deplored social injustice. He treated everyone, regardless of station, with dignity and courtesy and as an equal. A conversation with Howard-Flanders required patience. He spoke softly in a very deliberate manner. In fact, one was always tempted to complete his sentences because he would pause mid-sentence, think about what he was going to say, and after what seemed like an eternity, would finally complete the sentence. But his scientific insight was keen and he was a fierce competitor. The success of the radiobiology group at Yale was largely a result of the leadership of this remarkable man. His early death in 1991 left a large void in the hearts of his friends and colleagues.

Fresh out of his undergraduate studies at the University of Utah and hungry to engage in research "that would involve DNA" Boyce elected to pursue his graduate work with Dick Setlow in the Biophysics Department at Yale. He was particularly impressed by the enormous monochromator that Setlow had set up. "The entire monochromator was surrounded by black cloth and occupied space equivalent to a small bathroom," Boyce wrote. "In my mind it was a glorious creation and I greatly admired Setlow's ability to set up the optical and electrical components necessary for the successful operation of the monochromator." One of the amusing stories he related to me about life in the Setlow laboratory concerned a laboratory colleague who suffered from cold sores of the lip, which were greatly exacerbated by incidental exposure to low doses of UV light. The story has it that Setlow (unsuccessfully) pleaded with this individual to let him determine the UV action spectrum for the induction of cold sores!

Boyce's Ph.D. thesis centered on the mechanism whereby the thymine analog 5-bromouracil (5-BU) sensitizes cells to the lethal action of UV light. A primary conclusion that he drew from these studies was that the increase in UV sensitivity of 5-BU-substituted DNA in vivo may arise in part from the impairment of some sort of enzyme-catalyzed reactivation process. "Thus, the idea that radiation sensitivity could have a genetic component and that radiation damage might be enzymatically repaired was very much in my mind when I completed my doctorate in August of 1961—two years after arriving in New Haven," Boyce recalled.

The Howard-Flanders and Setlow laboratories enjoyed considerable informal communication facilitated by somewhat unique circumstances. Jane Setlow, Dick's wife at the time, had joined Howard-Flanders's laboratory in the

Department of Radiology at Yale as a postdoctoral fellow soon after obtaining her Ph.D. in the Biophysics Department. (Parenthetically, Jane Setlow and Phil Hanawalt were fellow graduate students and "lab partners" in the biophysics program at Yale). When Dick Setlow moved to Oak Ridge, Jane naturally followed him there and Dick Boyce replaced her as a postdoctoral fellow in Howard-Flanders's laboratory. So, Jane Setlow, a scientific collaborator of Dick Setlow, had worked with Howard-Flanders and liked him well enough that she continued to communicate with him and members of his laboratory on a casual basis. This sociology was complicated by the fact that Boyce, once a graduate student with Dick Setlow, remained on friendly terms with both Dick and Jane when he in turn moved to Howard-Flanders's laboratory. It does not take much leap of the imagination to anticipate a flash point in such a dynamic if the Setlow and Howard-Flanders laboratories worked competitively on closely related research topics, as indeed they did.

Like others interested in understanding the biological effects of UV radiation on bacteria, Howard-Flanders and Boyce were intrigued by the observation of Ruth Hill and her colleagues that the mutant that she had inadvertently isolated and called *E. coli* B_{s-1} was highly sensitive to UV radiation, and was additionally unable to support the growth of UV-irradiated phage T1, although unirradiated T1 phage grew perfectly well in this mutant. These results suggested to the Howard-Flanders group, as they had to Setlow, that "a factor capable of reactivating some sort of DNA photodamage was present in *E. coli* strain B but absent in strain B_{s-1}," wrote Boyce. Boyce and Howard-Flanders wanted to identify the putative gene(s) involved in this reactivation process by isolating their own UV-radiation-sensitive mutants. On the advice of Edward Adelberg, chair of the Microbiology Department at Yale, they elected to pursue their studies with the K12 strain of *E. coli*, because it was generally acknowledged to be genetically better characterized than the B strain. Boyce wrote:

> Paul's idea was to mutagenize the bacteria and plate them out on agar plates that had been spread with irradiated T1 phage. He reasoned that normal bacteria in the mutagenized population that were infected with the irradiated T1 would reactivate the DNA damage in the phage genome produced by previous exposure to UV radiation. Hence, the phage would replicate and destroy the normal bacteria. Mutant UV-sensitive bacteria would also be infected by the irradiated T1 phage, but in this case the UV damage would not be reactivated and the phage would not replicate in the mutant. Thus, colonies resistant to the UV-irradiated phage would appear after incubation of the plates.

This elegant screening strategy met their expectations handsomely and resulted in the isolation of the now famous *E. coli* K12 *uvrA*, *uvrB*, and *uvrC* mutants, which ultimately provided the means for identifying, purifying, and characterizing the UvrA, UvrB, and UvrC proteins, the essential primary components of the nucleotide excision repair machinery of *E. coli*.

Concurrent with these experiments, Boyce was trying to develop methods that would allow the identification and quantitation of thymine dimers, the suspected substrate for the "dark reactivation" process he had postulated. Like Carrier and Setlow, he radiolabeled the thymine in his *E. coli* parental and mutant strains and measured the dimer content of UV-irradiated cells by chromatographically resolving thymine and thymine dimers in acid hydrolysates of DNA. "After many attempts to optimize the system we obtained evidence that the dimers were disappearing from the DNA of the parental strain during post-irradiation incubation," Boyce communicated.

The history of events beyond this point is controversial. This controversy has been addressed in Carol Kahn's book entitled *Beyond the Helix: DNA and the Quest for Longevity*, to which the interested reader is referred for an independent accounting. Having had the opportunity of discussing this issue directly with Dick Setlow, Dick Boyce, and others, I thought it might be instructional to revisit this controversy. Dick Setlow's view is that having observed the disappearance of thymine dimers from high-molecular-weight acid-precipitable DNA, Boyce and Howard-Flanders were confounded by the fate of these dimers. Setlow recalls:

> Chromatography of the acid-soluble fraction of UV-resistant cells that had been UV-irradiated and incubated failed to show the presence of thymine dimers in their hands. Hence, Boyce and Howard-Flanders were unable to definitively determine the essential steps in the loss of dimers from the DNA of resistant cells—it could have transpired by excision repair or it could have happened by some type of direct dimer reversal, conceivably different from photoreactivation. We knew that the acid-soluble fraction, if it did contain dimers, would not have them in the form of free dimers but in the form of short oligonucleotides. We also knew that such oligonucleotides would not migrate chromatographically like dimers because of their length and because of the charge endowed by the phosphate groups on them. We indeed observed such small oligonucleotides in the acid-soluble fraction, but most importantly we showed that the acid-soluble fraction had to be hydrolyzed in acid in order for the thymine dimers to be observed chromatographically. Paul called Jane [Setlow] and asked how we could possibly find dimers in the acid-soluble fraction since they were not able to do this. He was told that all they had to do was to hydrolyze the acid-soluble fraction in acid. They did so and I believe that they then observed dimers.

Boyce has a somewhat different memory of the events at that time. He wrote to me:

> What I recall is that Jane paid us a visit at Yale during the time I was still setting up the paper chromatography and strip scanner to quantitate dimers in DNA. I vividly remember her visit because she stayed with us in our house in Orange and the hot water heater decided to go on the fritz! Our laboratories were still in the basement of the Hunter Radiation Building, and therefore this was well before we started to look for dimers in bacterial DNA. We performed all those experiments in our new laboratories in Sterling Hall some time later. She asked me the direction of our experiments and I told her that we planned to see if thymine dimers were "reactivated" in the uvr^+ [excision repair-proficient] but not in the uvr^- [excision repair-defective] K12 strains. Her reply was so odd that it has stuck in my mind. She said, "Why bother to do the experiment—surely you know that they [the dimers] will be reactivated." It is my contention that the Setlows knew at this point, well in advance of the time that Dick Setlow now claims that they found out through the [above mentioned] telephone call, that we were working on the problem. Moreover, Jane said not a word that they were also looking at the fate of dimers in Ruth Hill's strains.

Boyce elaborated further that:

> As I recall later events, Paul returned from a meeting at which he had heard Dick Setlow speak and he told me that Setlow had claimed that dimers were conserved

in intact *E. coli* B/r cells after UV-irradiation. We reasoned that if our experiments showed that dimers were disappearing from the DNA, but Setlow's experiments showed that dimers were conserved in intact cells, it could mean only one thing: they were being cut out of the DNA. Because we could not detect dimers in the incubation medium, it meant that they must trapped inside the cell. It was at this time that Paul suggested that we change our procedure to use cold trichloroacetic acid [TCA] to separate the bacteria into TCA-soluble and TCA-insoluble fractions. Whether that idea came to him through discussions with the Setlows, I can't say. The point was that the DNA and proteins should be present in the insoluble fraction, and all other cellular components, including thymine dimers, should appear in the soluble fraction. Thus, on July 26, 1963 we started to use the TCA treatment. But when I subjected the TCA-soluble fraction to paper chromatography no radioactivity appeared at the free dimer position. However, there was significant radioactivity that remained near the origin. It was at this point that I wrote or called Setlow and told him about our dilemma. He then gave us the valuable suggestion to hydrolyze the TCA-soluble fraction in hot acid prior to chromatography. This was the key to the experiment. We found on August 21, 1963 for the first time that dimers had been released from the DNA, presumably in a form that could not be transported out of the cell. We also recognized a technical advantage of looking for the dimers in the TCA-soluble fraction: the background level was lower and the results were clearer. We could now show that whereas the dimers are released into the TCA-soluble fraction in the uvr^+ strain, they were not released in the uvr^- strain. Shortly after this experiment, I informed Setlow of our success, and on September 23, 1963, he wrote me a letter (that I still have) in which his first sentence read as follows: "I am glad that you have found the dimers in the TCA soluble fraction of your [UV radiation] resistant bugs but not in the sensitive ones."

Another version of the revelation to the Howard-Flanders laboratory about what was transpiring in the Setlow laboratory vis-à-vis excision repair in *E. coli* was related to me by Evelyn Waldstein, a former Russian scientist now residing and working at the University of Tel Aviv in Israel. Evelyn first met Dick and Jane Setlow in 1972 when they visited Russia to attend an International Congress on Biophysics. Waldstein's introduction to the Setlows rapidly matured into a warm and comfortable friendship and culminated in an open invitation from Dick to Evelyn to visit his laboratory any time that she could manage to leave Russia, which Waldstein had confided to the Setlows she was passionately eager to do. Several years after this meeting, Waldstein was permitted to emigrate from the Soviet Union and in 1974 she spent a brief stint in Setlow's laboratory in Oak Ridge. As a devotee of the budding literature on DNA repair, a topic of her own scientific interest, Waldstein told me that she had always been "intrigued to know how two groups in the same country could have done essentially the same experiments and published them in the same issue of PNAS." So when she eventually arrived in Oak Ridge, a decade after these events, she specifically asked Bill Carrier about this. "He was still furious," recalled Waldstein, "and asked me 'Don't you know how this happened?'" Carrier communicated to her that around the time of the key experiments in both laboratories, Jane Setlow attended a meeting where she met Emmanuel Riklis (an Israeli post-doctoral fellow working in Howard-Flanders's laboratory). Carrier was of the view that Jane may have communicated information to Riklis about experiments in the Setlow laboratory and that it was very soon after this that Dick Setlow received a phone call from a perplexed Howard-Flanders informing him that they could not identify thymine dimers in the acid-soluble fraction of cells that had been exposed to UV radiation and incubated in the dark.

It is evident that both the Setlow and Howard-Flanders laboratories were independently heading in similar experimental directions. There is little question in the minds of most of the principal players in this scenario that Setlow and Carrier formally discovered excision repair by demonstrating that thymine dimers that were lost from high-molecular-weight DNA could be recovered in a fraction where they were presumed to be in small oligonucleotide fragments. There is also little question that Setlow provided significant technical help in facilitating a similar observation in Howard-Flanders's laboratory at about the same time. History will probably never reveal whether or not Boyce and Howard-Flanders had come to the independent conclusion that thymine dimers are excised from DNA based solely on analysis of their own data. Regardless, this series of events marred Setlow's future relationship with the Yale group. Setlow confided to me that ultimately what alienated him most was not that Howard-Flanders had exploited his technical advice, which was freely given, but the fact that soon after this Howard-Flanders contacted him and urged that the two laboratories publish their results as back-to-back communications in the *Proceedings of the National Academy* (*PNAS*). Setlow was offended by this suggestion, since it insinuated to him that Howard-Flanders was eager to claim fully equal credit for discovering excision repair in *E. coli*. Whether or not this was Howard-Flanders's real intention, and if so, whether this intention was legitimately founded, are nuances of history that will likely also never be resolved. Boyce of course was trapped in the middle of this delicate situation, as junior colleagues often are. He told me that he delayed writing up his results for publication for some time in the hope that Setlow would send him a preprint of his and Carrier's findings. When this was apparently not forthcoming, Howard-Flanders asked Max Delbrück to communicate their results to *PNAS*. "His response was immediate, favorable, and surprising," Boyce informed me. "Delbrück said that he had recently seen a manuscript from Setlow and Carrier describing similar results. He felt that this was an exciting and important discovery, and that both papers should be published in the same volume of *PNAS*." From my correspondence with Dick Boyce, it was evident that he was pained by the nature and persistence of the controversy surrounding the discovery of excision repair in *E. coli*. He wrote to me that:

> It is probably beyond reason to expect [Dick] Setlow to concede that each of our excision papers complemented and enhanced the other. However, it is my perception that this was exactly Delbrück's motivation when he saw to it that both papers were published simultaneously. I think he saw the significance and importance of this discovery far more clearly than we did at the time, and that as a true scientist he was less interested in who won the glory than in getting the corroborated results to the scientific community.

The papers by Carrier and Setlow and by Boyce and Howard-Flanders were published within 60 pages of each other in the 1964 volume of *PNAS*. Boyce informed me that in 1979 when their *PNAS* paper was identified for the so-called "citation classic" category of the weekly periodical *Current Contents*, he was asked by Susan Perretta of *Current Contents* to write a short essay on the events surrounding this work. He confided:

In that essay I acknowledged the technical breakthrough which Setlow's suggestion made possible. Furthermore, I stated that their work predated that of our own, and paid tribute to the significance of their work. I welcome this opportunity to try to set the record straight based on my best recollections, backed up with some documentation. I continue to respect Setlow as both a scientist and as a person. I acknowledge his very fine contributions to the field of DNA repair and wish him many productive years.

To conclude this interesting historical anecdote, I believe it fair to point out that when I asked Dick Boyce to comment on this chapter prior to publication he expressed his profound disappointment that my treatment of the discovery of excision repair was heavily biased in favor of the roles of Setlow and Hanawalt.

The discoveries by Setlow and Howard-Flanders and their colleagues were featured at a symposium of the 8th Annual Meeting of the Biophysical Society in Chicago, in February 1964. The full implications of the newly-discovered excision repair pathway in *E. coli* were extensively reviewed at a more specialized meeting convened in Chicago in October 1965 under the title *Structural Defects in DNA and Their Repair in Microorganisms*. This meeting, organized by Bob Haynes, Sheldon Wolff, and James Till, is a notable historical landmark in the DNA repair field, since it represents the first meeting formally dedicated to this rapidly emerging discipline. The 43 attendees included Dick and Jane Setlow, Bill Carrier, Paul Howard-Flanders, Dick Boyce, Reginald Deering, Alex Hollaender, Stan Rupert, Ruth Hill, Henry Kaplan, Walter Harm, Phil Hanawalt, Larry Grossman, Bernard Strauss, Evelyn Witkin, and fittingly, Max Delbrück, who chaired the concluding summary session. This meeting not only consolidated progress in the repair of base damage (primarily photoproducts) in DNA, but also provided an important synthesis between excision repair and UV-induced mutagenesis. In the published proceedings of the meeting, Evelyn Witkin provided the important emphasis that:

> There is probably no real distinction between "lethal" and "mutagenic" UV photoproducts. A bacterium survives irradiation if (a) the residual damage (after repair) does not exceed a certain strain-specific limit still compatible with DNA replication, and (b) at least one DNA strand is free of induced lethal mutations. Such a cell will produce at least one viable segregant. An induced mutation in a particular gene will be obtained from such a survivor if an unrepaired UV lesion happens to be located in the gene in question, and if it causes a replication error during DNA synthesis.

Of course, at that time no one could predict just how complex the process of mutagenesis induced by UV radiation was. The unfolding of this complexity will be visited in detail in a later chapter.

The year 1964 was a banner one for excision repair. Immediately following the reports by Setlow and Carrier, Boyce and Howard-Flanders and Pettijohn and Hanawalt, delineating the essential features of nucleotide loss and replacement that characterize all forms of this DNA repair mode, preliminary evidence suggesting that this type of repair also operates in mammalian cells was announced by Robert Painter and his colleague Ronald Rasmussen, then at the NASA Ames Research Laboratories near Palo Alto, California. Painter and Rasmussen were using autoradiography of cells labeled with tritiated thymidine to investigate the effects of ionizing radiation on DNA synthesis in mammalian cells. Having caught wind of the recent discovery of excision repair in bacteria they were prompted to examine the pattern of DNA synthesis in mammalian cells exposed to UV light. Years later, Painter wrote that:

> What we found was... [a] shock. Not only were all S-phase cells participating, but *all* the cells in the culture were synthesizing DNA!

Painter and Rasmussen's observations strongly hinted that repair synthesis, and hence presumably excision repair, was an evolutionarily conserved mechanism. However, other explanations were clearly tenable, including the possibility of aberrant reinitiation of semiconservative DNA synthesis as a result of exposure of mammalian cells to UV radiation. In the mid nineteen-sixties, eukaryotic molecular biology was still in its infancy and the use of more direct techniques for demonstrating excision repair in mammalian cells, such as measuring the thymine dimer content of the acid-insoluble and acid-soluble fractions of cells, were still flawed by technical problems not encountered in *E. coli*. Setlow drew attention to these problems at the 1965 Chicago meeting on DNA repair alluded to above:

> Excision... has been looked for and *not* found in mammalian cells (Chinese hamster and kidney) in tissue culture. The DNA of such cells is organized differently from that of bacterial cells, and it is possible that the excised pieces are large enough to be acid-insoluble. More sophisticated DNA-fractionation procedures must be utilized, and cell lines derived from tissues that are normally sensitive to UV must be investigated before we reach any firm conclusions for mammalian cells.

Painter and Rasmussen experienced similar difficulties. In their 1964 paper, in which they suggested repair synthesis in HeLa cells, they stated that:

> The results reported here may involve a system similar to that reported for bacteria by Setlow and Carrier and Boyce and Howard-Flanders. In the case of bacteria the effect was detected by the finding that thymine dimers moved from the acid-insoluble to the acid-soluble fraction in the first hour after ultra-violet irradiation. We have attempted to determine, by pre-labeling the DNA with ^{3}H TdR and ^{14}C BUdR, if thymine or BU in the DNA of cells irradiated with ultra-violet light is lost to the acid-soluble fraction (or to medium, wash fluid, etc.). So far we have found no quantitative increase in material lost from DNA of irradiated cultures over that in controls.

Perhaps for these reasons the observation of putative repair synthesis in human cells went largely unnoticed for a while. Painter recalled:

> Nobody wrote and nobody called to ask about or to comment on these results. Moreover, not another paper was published to corroborate our results for over two years, and that one, by Djordjević and Tolmach, did not appear until after our second paper on the subject. This loud silence was disturbing and for a long time I thought perhaps we were somehow the only people in the world who could obtain these "strange" results. In early 1966 I attended the annual meeting of the Radiation Research Society in Coronado. I noted from the abstracts that no one else was presenting anything on radiation-induced DNA synthesis and I still had this feeling that maybe we were doing something wrong. The meeting took place right at the height of one of the Asian flu epidemics and my roommate (Ted Phillips) and I spent much of the time in bed with the miseries. On the penultimate day of the meeting I began to feel better and went down to the lobby. I sat next to Ron

Humphrey, whom I hardly knew at that time, and we began a conversation. He offhandedly mentioned that he and his colleagues had observed radiation-induced DNA synthesis in several mammalian cell lines! I was elated and told him how I had been afraid that we were somehow wrong.

The putative repair synthesis observed by Rasmussen and Painter in human cells by simple autoradiography of UV-irradiated cells was termed "unscheduled DNA synthesis" (UDS) by B. Djordjević and L. J. Tolmach, who confirmed Painter's observations. The term refers to the fact that DNA synthesis was occurring outside of the scheduled S phase of the cell cycle. The NASA laboratories were conveniently located to Stanford, and when Painter learned that David Pettijohn and Phil Hanawalt had a more definitive method for demonstrating repair synthesis of DNA, he journeyed up the south bay. In due course, Painter confirmed, with Hanawalt's assistance, that the autoradiographic grains observed in UV-irradiated mammalian cells did indeed represent a form of non-semiconservative DNA synthesis—presumably, the repair synthesis associated with excision repair. Subsequently, the excision of dimers was directly demonstrated in mammalian cells in several laboratories.

In 1965, Painter was recruited by Harvey Patt to the Laboratory of Radiobiology, a new National Laboratory supported by the Atomic Energy Commission on the San Francisco campus of the University of California. Patt hoped to assemble a first class team of senior investigators to engage in state-of-the-art radiobiological research, each of whom would have the opportunity to recruit a junior associate of their choice. Painter was very keen to recruit a young Englishman named James Cleaver, who had completed his Ph.D. in physics at Cambridge on the effects of radiation on DNA synthesis, and who in 1965 was engaged in postdoctoral training in Boston. As a recent visitor to the United States, Cleaver wanted to explore more of the country and decided to motor down the East Coast from Boston to Florida. During this journey he stopped off at Oak Ridge to visit with Setlow, who in fact offered him a second postdoctoral position in his own laboratory. Soon thereafter Cleaver renewed his acquaintance with Painter at the Annual Biophysical Society meeting and received an offer from him as well. "I finished up with two offers, one in Tennessee and one in San Francisco," Cleaver told me. "They both came in the middle of winter, so it wasn't a difficult choice!" Cleaver joined the staff of the Laboratory of Radiobiology at the University of California, San Francisco in 1966 and has remained in sunny California ever since. It was here that he made his startling and pioneering observations of defective excision repair in human patients with the disease xeroderma pigmentosum (XP). Cleaver recounted:

> I was working with Painter shortly after he had discovered unscheduled DNA synthesis. We were trying to adapt Hanawalt's BrdU [bromodeoxyuridine] method to measure repair synthesis in mammalian cells. But the driving question was how to obtain UV-sensitive mammalian cell mutants to prove that the repair synthesis was really biologically related to excision repair. We were going to attempt to make such mutants ourselves. In April of 1967, I saw an article in the *San Francisco Chronicle* by the science writer David Perlman. It was a brief report of a clinical meeting—the 48th Annual Meeting of the American College of Physicians in San Francisco, and highlighted a talk on the genetics of human cancer by Henry Lynch. The article described xeroderma pigmentosum as a genetic disease with a

predisposition to skin cancer and sensitivity to sunlight. I thought to myself, "My word, here are God-given UV-sensitive mutants." So we worked through the dermatology department at UCSF and got skin biopsies and cultures of XP patients. We got three cultures from three different patients with XP. Painter and I used the methods we had developed at that time for unscheduled DNA synthesis and Hanawalt's technique for repair replication. The results came up on each cell line right away.

I became aware of Cleaver's fascinating observations that XP cells were defective in unscheduled DNA synthesis and in repair synthesis of UV-irradiated DNA after they were published in *Nature* in 1968. I must frankly admit that my immediate pleasure at this extraordinary link between DNA repair and human disease was tainted by an inescapable sense of jealousy that this was the "perfect" discovery for a pathologist interested in DNA repair; but it had been made by a damn physicist instead! In the 27 years since Cleaver's description of defective excision repair in XP patients, progress in our understanding of the molecular pathology of this disease has been nothing short of spectacular and will provide fertile material for future historians interested in this topic.

Cleaver shared with me the irony of how close he came to being hit by lightning twice in the same storm. Prior to the discovery that cells from individuals with the hereditary disease ataxia-telangiectasia (A-T) are X-ray sensitive—an observation that hinted at yet another relationship between DNA repair and human disease—such cells were, in fact, established in culture in Painter's laboratory. Cleaver selected these as a control for his XP studies simply because they were convenient:

> I decided to used some other randomly selected disease cell line as a control to show that the low labeling in the XP cells wasn't the non-specific result of a disease state. By the time I started a new series of experiments designed to examine X-ray repair in XP cells the AT line had died out!

Had this not been the unfortunate case, Cleaver almost certainly would have made the second major discovery that A-T cells were abnormally sensitive to killing by X rays. This significant fact was not revealed until several years later by Malcolm Taylor and his colleagues in the United Kingdom.

Bob Haynes chuckled when we talked about Cleaver's discovery of the excision repair defect in xeroderma pigmentosum. "I can vividly remember saying to several people, thank God, we've now got a disease. We'll be able to get money from NIH because DNA repair is relevant to a human disease!" In the same proselytizing vein, Joshua Lederberg, then Chairman of the Department of Genetics at Stanford University, wrote an editorial for the Washington *Post* in June 1968 soon after Cleaver published his observations:

> Excitement about the new biology of DNA has tended to provoke either of two reactions: that little men would soon be synthesized and come swarming out of the laboratories or that the whole study of molecular biology was mainly of academic importance and we would see little practical impact of it within our lifetime.
>
> More sensible observers have suggested that we would see ever deeper insights into cancer and other cell pathology over the coming years and decades. It is gratifying and a little surprising, to see how quickly a concrete application of DNA enzymology has arisen in relation to a human disease....
>
> [Cleaver's] work exemplifies the excellent biomedical research at our universi-

ties, supported by agencies like the Atomic Energy Commission, the National Institutes of Health and the National Science Foundation. Its further growth, however, is being strangled in the budgetary crunch between the President and Congress. It is no consolation that their children will share in paying the price of their neglect.

The history of scientific exploration is replete with evidence that when an idea is ripe it will frequently surface in more than one active mind at about the same time. Prior to joining the faculty at Rutgers University in 1971, Evelyn Witkin spent 15 years in the Department of Medicine at the Downstate Medical School of the State University of New York. Not being medically trained, Witkin's didactic teaching duties were largely confined to communicating new advances in genetics to third-year medical students in a few lectures each year. It has certainly been my own experience that senior medical students, saturated with the tedium of several years of basic science taught mainly in the classroom, and now eager to hone their skills as practicing clinicians in hospital wards, are not especially receptive to yet more lectures on basic science. It is also my distinct recollection that during the late nineteen-sixties, medical students were especially resistant to topics that they deemed "irrelevant" to the care of the ill, a theme that was, in their view, consonant with the culture of "relevance" very much in vogue at that time. In preparing her annual lectures to the students in 1967, Witkin made a special effort to interject as much clinical "relevance" as possible to her anticipated discussions on DNA repair, the topic of her choice. She therefore visited the library to explore human diseases that might conceivably qualify as being defective in this aspect of gene metabolism. "Among others," she told me, "I came up with XP right off the bat and showed the students slides of XP patients, suggesting that they might have a problem with excision repair of DNA. Imagine my delight when the Cleaver paper came out!"

A significant milestone in the history of the discovery of excision repair was the early recognition that in contrast to the absolute specificity of photoreactivation for pyrimidine dimers, this type of DNA repair was more general. As mentioned earlier, during the course of his extensive studies on liquid holding recovery of UV-irradiated cells in the early nineteen-sixties, Haynes and his collaborators extended their repertoire of DNA-damaging agents to include the alkylating agent nitrogen mustard, a drug recently introduced to the clinical arsenal of chemotherapeutic agents for the treatment of cancer. "The reason we were working with nitrogen mustard was that it was being used clinically as a so-called radiomimetic drug," related Haynes. "One of the things we did was to examine liquid holding recovery after nitrogen mustard exposure. When I looked at the sensitivities of $E.$ $coli$ B/r and B_{s-1}, the curves were incredibly similar to those with UV radiation."

In the spring of 1964, Haynes moved to the Donner Laboratory in the rolling hills of the University of California campus in Berkeley, another National Laboratory supported by the U.S. government. Here he renewed his long acquaintance with Phil Hanawalt at nearby Stanford, and was, of course, thoroughly familiar with the results of David Pettijohn's experiments. Both he and Hanawalt attended the International Congress of Photobiology in Oxford in 1964, where the subject of the survival curves of $E.$ $coli$ B/r and B_{s-1} treated with nitrogen mustard surfaced over drinks in an Oxford pub. Hanawalt later wrote that:

Bob Haynes and I set out to test the generality of excision repair beyond pyrimidine dimers, choosing to study the bifunctional alkylating agent nitrogen mustard. Haynes had shown that the respective survival curves for *E. coli* wild-type and mutant strains exposed to this agent strikingly paralleled those for UV. We observed repair replication in nitrogen mustard-treated cells....

The generality of excision repair was independently documented by Dick Boyce soon after he and Howard-Flanders published their results on excision repair of pyrimidine dimers. Boyce was aware that as early as 1961 Edward Reich and his colleagues had noted that when *E. coli* cells are exposed to mitomycin C their DNA undergoes degradation. Boyce wondered, "Could this breakdown be a reflection of excision repair of mitomycin C-induced damage in the DNA?" He decided to approach the problem by first examining the lethal effects of mitomycin C (MC) on his collection of excision-defective *uvr* strains. Boyce informed me:

If it appeared that *uvr*⁻ strains were unable to repair MC-induced damage, I planned to examine DNA breakdown in the same strains as an indirect manifestation of excision repair. When I discussed these ideas with Paul, he bet me two beers that the *uvr* genes would have no effect on MC-induced damage. When I showed him the survival curves a few days later he was flabbergasted. The lethal effects of UV and MC were identical. Neither agent produced damage that could be repaired in any of the *uvr*A, *uvr*B or *uvr*C mutant strains. We wrote this work up and submitted it to Delbrück. This time he felt that it wasn't a *PNAS* paper. He wrote back saying, "I think it has not been shown that DNA breakdown is an essential part of the excision mechanism and it has not been shown, in the case of mitomycin C, that repair synthesis has taken place." However, he thought that the manuscript "contains a number of observations of great interest," and offered to send it to *Z. Vererbungslehre*, of which he was an editor. We accepted his offer, and this paper was published in that relatively obscure journal in the fall of 1964.

Boyce and Howard-Flanders' paper presumably escaped the notice of Hanawalt and Haynes.

It is now firmly established that the generality of excision repair embraces a wide variety of chemicals, many of which are proven mutagens and carcinogens. But not all base damage is repaired by the excision repair mode discovered by Setlow. It required several new and independent discoveries to show that the generality of excision repair is contributed in part by several biochemically distinct pathways and modes. The history of these discoveries is the topic of the next chapter.

Once the phenomenon of excision repair was clearly established, the challenge was once again for the biochemists to identify and characterize the specific enzymes involved. This proved to be a daunting task. Part of the problem was the biochemical complexity of excision repair in *E. coli*, first hinted at by Boyce and Howard-Flanders's identification of three genetic loci. More devilish from the biochemist's point of view is the fact that like photoreactivating enzyme, the products of the *uvr*A, *uvr*B, and *uvr*C genes are constitutively expressed in tiny amounts in *E. coli*, and again it required the enormous amplifying power of recombinant DNA technology to solve this limitation. This solution was late in coming, and while its ultimate arrival ushered in a new golden era in the history of DNA repair, this period of discovery is relatively recent, and I will leave it for some future historian to recount in detail.

Hermann J. Muller, whose work on the effect of X rays on mutagenesis in *Drosophila* won him the Nobel Prize. *Courtesy of Cold Spring Harbor Laboratory Archives*

Milislav Demerec, former director of Cold Spring Harbor Laboratory. *Courtesy of Cold Spring Harbor Laboratory Archives*

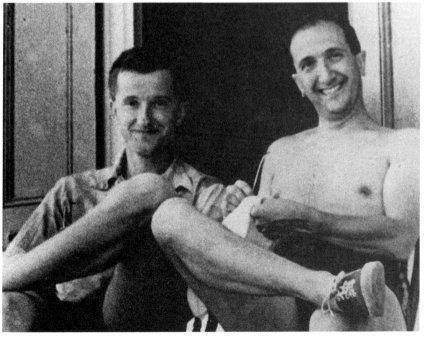

Max Delbrück (left) with Salvador Luria at Cold Spring Harbor. *Courtesy of Cold Spring Harbor Laboratory Archives*

Alexander Hollaender, a pioneer in the DNA repair field and for many years director of the Biology Division, Oak Ridge National Laboratory. *Courtesy of Environmental Mutagen Society Archives*

Albert Kelner discovered photoreactivation in the late nineteen-forties. *Courtesy of Adelyn Kelner*

Renato Dulbecco independently discovered photoreactivation while he was a postdoctoral fellow with Salvador Luria. *Courtesy of Dr. Renato Dulbecco*

Claud S. (Stan) Rupert discovered photoreactivating enzyme with his mentor Sol Goodgal. *Courtesy of Dr. Stan Rupert*

Sol Goodgal. *Courtesy of Dr. Sol Goodgal*

Evelyn Witkin contributed much to our current understanding of DNA repair and mutagenesis. *Courtesy of Dr. Evelyn Witkin*

Participants at the first formal conference on DNA repair sponsored by the National Academy of Sciences National Research Council of the United States of America in Chicago, October 18–20, 1965. Principals featured in this story include: *Front row, from left:* E.P. Geiduschek; David M. Freifelder; Richard B. Setlow; Stan Rupert; Ruth F. Hill; Evelyn M. Witkin. *Second row, from left:* Jane K. Setlow; Lawrence Grossman; fifth from left, Walter Harm; far right, Richard Boyce. *Third row:* Fifth from left, Philip C. Hanawalt. *Fourth row:* Second from left, Max Delbrück; third from right, Arthur Rörsch; extreme right, Bernard S. Strauss. *Back row:* Third from left, Paul Howard-Flanders; fourth from left, Henry S. Kaplan; fifth from left, Robert Haynes.

From left to right, John Jagger, Robert H. Haynes, Sheldon Wolff, and Philip C. Hanawalt at the Gatlinburg Symposium, Tennessee, 1967. *Courtesy of Dr. Robert Haynes*

Richard B. Setlow discovered excision repair of DNA.
Courtesy of Dr. Richard Setlow

Paul Howard-Flanders, together with Richard Boyce, independently discovered excision repair.
Courtesy of June D. Howard-Flanders

Richard Boyce. *Courtesy of Dr. Richard Boyce*

Philip Hanawalt (left) with Robert Haynes in St. Albans, England, on the occasion of their 62nd birthdays in 1993. *Courtesy of Dr. Philip Hanawalt*

Bernard Strauss (left front) with James E. Cleaver (right front), Erling Seeberg (left rear), and the author (rear), 1982. *Courtesy of Dr. Bernard Strauss*

Bernard Strauss with Mutsuo Sekiguchi in Japan, 1981. *Courtesy of Dr. Bernard Strauss*

Tomas Lindahl (shown here with two of his postdoctoral fellows) discovered base excision repair. *Courtesy of Dr. Tomas Lindahl*

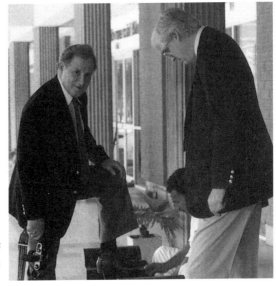

Lawrence Grossman with Hamilton Smith of Johns Hopkins University in Pakistan, circa 1984.

John Cairns with Leona Samson at Cold Spring Harbor, June 1978. *Courtesy of Dr. Leona Samson*

Sanford Lacks. *Courtesy of Dr. Sanford Lacks*

Matthew Meselson identified the role of DNA methylation in strand discrimination during mismatch repair.
Courtesy of Dr. Matthew Meselson

Miroslav Radman in 1981.

Raymond Devoret, circa 1969.
Courtesy of Dr. Raymond Devoret

CHAPTER 4
New Mechanisms for Repairing DNA

THE DISCOVERY OF EXCISION REPAIR by Dick Setlow and Paul Howard-Flanders and their colleagues prompted a number of investigators to go in search of the enzymes responsible for this metabolic transaction in damaged DNA. This chapter begins with an account of the tortuous history of some of the principal efforts in this regard and their unwitting (and for many years unrecognized) inroads into an entirely new mechanism for the excision repair of damaged DNA.

When I finally moved to Stanford in 1971 and shed the uniform of an officer of the U.S. Army, I continued my studies on the enzyme activity from phage T4-infected cells that Jack King and I had discovered at the Walter Reed Army Institute of Research. By this time, I knew that the enzyme was an endonuclease that uniquely and specifically attacked UV-irradiated DNA. It later turned out that this specificity provided great utility for the sensitive detection of pyrimidine dimers in DNA. However, at the time I was frankly disappointed because the generality of excision repair for many different types of base damage in *E. coli* was now well recognized, and the specificity of the phage-encoded enzyme for pyrimidine dimers strongly suggested that phage T4 would not to be an informative model for the mechanism of excision repair in *E. coli*. I was having no success in identifying a DNA damage-specific endonuclease in extracts of uninfected *E. coli*, so I decided to pursue the characterization of the T4 enzyme and to elucidate the biochemical pathway in which it operated. For want of a better name, I dubbed this enzyme the ultraviolet (UV) endonuclease of T4. It was independently called endonuclease V (five) by Mutsuo Sekiguchi, but ultimately achieved the abbreviated designation denV endonuclease of phage T4, after the *denV* (*d*NA *e*ndonuclease V) gene (formally the v^+ gene) which encodes it. I once submitted a manuscript to the *Journal of Biological Chemistry*, the title of which included the phrase "ultraviolet (UV) endonuclease." I received an irate letter from the Editor-in-Chief, Herbert Tabor, informing me that since the protein was presumably not colored ultraviolet the name was blatantly ridiculous! I gather that many other contributors to the journal have been on the receiving end of Dr. Tabor's terminological austerity.

As I mentioned in the previous chapter, the phage T4 repair enzyme was anticipated by Dick Setlow and Bill Carrier as early as 1966 and was independently discovered by Mutsuo Sekiguchi and his colleagues in Japan. Mutsuo Sekiguchi has been an esteemed colleague and gracious friend for many years. During one of my several visits to Japan, I had the memorable experience of staying with him for a few days in a most elegant Japanese inn, access to which is not usually available to unaccompanied Westerners. I sometimes still wonder how badly I embarrassed Mutsuo by my limited familiarity with Japanese etiquette during that stay. An equally memorable experience was dining on fugu with him. Mutsuo insisted that I defer to his tasting the fish first. Much to my relief, he did not manifest any noticeable immediate effects. However, as I began my own consumption of this truly exquisite offering, I could not help wondering what the latent period was.

Following the completion of his graduate studies, Sekiguchi, like many of the best and brightest young scientists in Japan, journeyed to the United States to pursue postdoctoral training, in his case with Seymour Benzer at Purdue University and then with Seymour Cohen at the University of Pennsylvania. He returned to Japan in 1965 where he was appointed to a faculty position at Kyushu University on the beautiful island of Kyushu. He told me:

> I was excited about getting ready to start my own research projects in Japan. But I also experienced moments of intense frustration because even in the mid nineteen-sixties there was still a tremendous gap between the high level of academic institutions in the United States and the postwar problems lingering at universities in Japan. Research funds were in short supply and even basic equipment was sparse.

These limitations influenced Professor Yasuyuki Takagi, the head of the research group of which Sekiguchi became a new member, "not to tackle problems as complicated as DNA replication," Sekiguchi related. "Studies in nucleic acid synthesis were ongoing in the United States and many formidable groups were emerging. To enter this particular field was like trying to sweep the sea with a broom!"

Sekiguchi and his colleagues were aware of the recent exciting work from Dick Setlow's and Paul Howard-Flanders's laboratories, and like me (and, as we shall presently see, several other biochemists), they were keen to identify and characterize the enzymes responsible for excision repair in *E. coli*. Takagi's group reasonably concluded that an enzyme that excises pyrimidine dimers was likely to be a nuclease specifically active on UV-irradiated DNA. Hence, potential assays for detecting such an activity were quickly devised. The Japanese team included Shunzo Okubo, who had spent two years working in Bernie Strauss's laboratory at the University of Chicago as a postdoctoral fellow. Having visited Japan in 1956 as part of a formal U.S. delegation of scientists invited to participate in one of the earliest postwar scientific meetings held in that country, Strauss had become something of a Nipponophile and had returned to Japan for a sabbatical in 1958. Strauss told me:

> I had already made arrangements with Mogens Westergaard to go to his laboratories in Copenhagen. However, the appeal of Japan was just too great. I wanted to see more of that country even though it was quite clear that for professional advancement I should probably go to Europe.

Strauss spent his sabbatical in Hideo Kikkawa's laboratory at Osaka University, where he first met Mutsuo Sekiguchi who was one of Kikkawa's graduate students. As a consequence of the extensive rapport between Strauss and the Japanese investigators, Sekiguchi and others were well aware that as early as 1962, Strauss had demonstrated that extracts of the highly UV-radiation-resistant bacterium, *Micrococcus lysodeikticus* (now called *Micrococcus luteus*), inactivated transforming DNA that had been exposed to UV light. I shall return to the specifics of these observations by Strauss and his colleagues shortly.

So, the Japanese group initially focused their efforts on extracts of *M. lysodeikticus*. They observed preferential degradation of UV-irradiated DNA as well as the specific excision of pyrimidine dimers in vitro. Convinced that these findings paralleled the (anticipated) enzymology of excision repair in *E. coli*, Sekiguchi and his collaborators eagerly sought mutants of *M. lysodeikticus* that were defective in this enzymatic activity, but without success. Sekiguchi ruefully reflected that:

> The going was very slow so we eventually decided to settle on bacteriophages, which have a simpler genetic system. At that time, it was known that some T4 phage mutants were sensitive to UV light. Moreover, this approach was advantageous to me as my postdoctoral work was mainly concerned with phage biochemistry and genetics. We soon found that an extract of phage T4-infected cells contained a dimer-releasing enzyme activity, whereas an extract of UV-sensitive T4$_{v-1}$-infected cells did not.

This work was presented by Takagi in the context of a larger study at the 1968 Cold Spring Harbor meeting on *Replication of DNA in Microorganisms*, a meeting that I attended as a young postdoctoral fellow. But I have no recollection of Takagi's talk since, aside from the fact that I was totally overwhelmed by the experience of attending my first international scientific meeting, I spent most of it in a state of intense anxiety at the prospect of having to present the results of my own recent work in David Goldthwait's laboratory at Case Western Reserve University, some of which I shall recount shortly. The results of the Takagi group were documented in the published proceedings of the Cold Spring Harbor meeting. By the time this volume was published, Jack King and I had carried out similar experiments at Walter Reed. But the Japanese publication preceded ours in 1969 and we dutifully acknowledged it. So, both Mutsuo Sekiguchi and his colleagues in Japan and Jack King and I working at Walter Reed stumbled onto the T4 pyrimidine-dimer-specific enzyme for the same well-intentioned but misguided reasons.

Although he did not pursue his initial biochemical observations in any detail, history must credit Bernie Strauss with the discovery of the pyrimidine-dimer-specific activity of *M. lysodeikticus*, later shown to be a catalytic function of an enzyme with properties identical to those of the T4 enzyme. As just recounted, it was in fact this enzyme that temporarily lulled Sekiguchi and his coworkers into thinking that the activity they found in extracts of *M. lysodeikticus* also operated in *E. coli*. Strauss was introduced to *M. lysodeikticus* when he joined the faculty of the Department of Microbiology at the University of Chicago in 1960. He was interested in understanding the mechanism of genetic recombination and he and his colleague Peter Guiduschek, also at the University of Chicago, set about trying to demonstrate recombination between *B. subtilis* and *M. lysodeikticus* DNA. They selected these two genomes because of the large difference in their buoyant densities, anticipating that recombinant DNA molecules would be readily detected by a density shift. Strauss related:

> I had both B. subtilis and M. lysodeikticus in the laboratory. And with Peter Guiduschek's lead, I was mixing their DNA together with cell extracts and analyzing the products of these reactions using a DNA transformation assay. Meanwhile, I was becoming acquainted with the experiments going on upstairs by Bob Haynes and his graduate student Michael Patrick, and Bob Uretz. They were irradiating yeast and then storing cells in buffer for periods before allowing growth, and were discovering that there was some sort of recovery process. That was when I first became aware that some people thought that DNA could actually be repaired! I was also studying the recovery of bacteria after treatment with alkylating agents. While riding on the Illinois Central one time, it occurred to me that our experiments indicated that the transforming activity of alkylated DNA was reduced a lot more than that of control DNA after treatment with cell extracts. This suggested some sort of enzymatic degradation process that was specific for alkylated DNA, but I resisted the idea that this has anything to do with repair. I am even less sure why I also treated UV-irradiated DNA with the cell extracts, but I eventually realized that the M. lysodeikticus extracts reduced the activity of both UV-irradiated and alkylated DNA, whereas the B. subtilis extracts reduced the activity of only the alkylated DNA.

The repair of UV radiation damage that Strauss observed in extracts of M. lysodeikticus and that of alkylation damage that he simultaneously noted in extracts of B. subtilis use distinct enzymes. But, both forms of repair occur by the same general excision repair pathway. As we will presently see, it turned out that this pathway (now formally designated as *base excision repair*) also incorporates that involving the phage T4 v^+ gene product that Sekiguchi and I independently discovered, and is quite different from that used for the repair of UV radiation damage in E. coli (now designated as *nucleotide excision repair*).

Yet another important contributor to the DNA repair field who was intent on deciphering the enzymology of excision repair in E. coli fell into the (almost irresistable) trap of using M. lysodeikticus or phage T4-infected E. coli as model systems because all these systems shared in common the specific degradation of UV-irradiated DNA. At about the same time that Sekiguchi and his colleagues in Japan and Jack King and I in Washington D.C. were probing the biochemistry of excision repair in T4-infected cells, the enzymatic activity identified in extracts of M. lysodeikticus by Strauss in 1962 was extensively purified by Lawrence Grossman and his colleagues at Brandeis University. Grossman traces his interest in DNA repair to a seminar from S. Y. (Sidney) Wang that he heard when he (Grossman) was a young faculty member at Brandeis University. As recounted in Chapter 2, Wang had at that time independently discovered that pyrimidine dimers are formed in UV-irradiated frozen solutions of pyrimidines. Little did Grossman realize at this early stage of his career that his budding interest in DNA damage and repair would eventually lead to his succession to the Chair held by Roger Herriott in the Department of Biochemistry at the Johns Hopkins University School of Hygiene and Public Health, the very department that spawned one of the major pioneering efforts in the history of the DNA repair field—the discovery of photoreactivating enzyme by Sol Goodgal and Stan Rupert (see Chapter 2). Together with his faculty colleague, Julius Marmur, Grossman initiated extensive studies on the photochemistry of nucleic acids. Among his several contributions in this arena, was the observation that one of the stereoisomers of thymine dimers could covalently link the two polynucleotide chains in UV-irradiated DNA, forming DNA cross-links. "This proved to be readily demonstrable photochemically," he told me, "but was unlikely to be biologically significant because of its low quantum yield." Nevertheless, Grossman's photobiological investigations attracted the interest of Max Delbrück, who invited him to spend the summer of 1961 at his new research institute in Cologne. Like many others who crossed Delbrück's path, Grossman was left with indelible impressions of the man. "Working with Max figured very importantly in my scientific life and my perceptions of the scientific ethic," he told me. In particular, Grossman recollects sending Delbrück a proposed abstract for a meeting based on work carried out during that summer in Cologne. The abstract (legitimately) identified Delbrück as a coauthor, but Delbrück insisted that it was unjustified for him to be included since he had neither "participated in the work nor labored over a single word of the writing."

As was the case with the other research groups noted above, the discoveries by the Setlow and Howard-Flanders laboratories inspired Grossman to pursue the protein factors responsible for excision repair of photochemical damage. "In our search for damage-specific enzymes, we sought a UV-radiation-resistant organism possessing low endogenous levels of nonspecific nucleases and settled on studying excision repair in *Micrococcus lysodeikticus* (*luteus*)," he recalled. Inevitably, Grossman encountered the pyrimidine-dimer-specific enzyme so abundantly present in extracts of this organism. He and his young colleague Robert Grafstrom in collaboration with William Haseltine and his colleagues at the Dana Farber Cancer Center in Boston ultimately elucidated the novel and rather unexpected mechanism of action of this enzyme in the late nineteen-sixties. An accounting of this mechanism is deferred until later in the chapter because when this secret was ultimately revealed, the *M. luteus* (and phage T4) enzymes were found to be unanticipated members of an already known class of DNA repair enzymes called DNA glycosylases, the discovery of which is one of the primary themes of this chapter.

Long before the precise mechanism of action of the *M. luteus* and phage T4 enzymes was understood, several investigators had begun cleverly to exploit the stringent substrate specificity of these enzymes in studies designed to show that pyrimidine dimers were responsible for one biological effect or another. Among them was a research associate with Phil Hanawalt at Stanford University, Ann Ganesan. Our close physical proximity at Stanford provided Ann ready access to T4 UV endonuclease purified in our laboratory. Dick Setlow and others were using the *M. luteus* enzyme as a dimer-specific reagent with similar success. Photoreactivating enzyme, also strictly specific for pyrimidine dimers, would have been equally well suited for such purposes, but as mentioned earlier, this enzyme was difficult to purify because of its very low abundance in cells and hence it was not easily available as a dimer-specific probe. The phage T4 and *M. luteus* enzymes shared another very attractive feature. Neither required divalent cations as cofactors, as many nucleases do, and hence they were fully active in the presence of chelating agents. Since many of the nonspecific nucleases in extracts of T4-infected *E.*

coli and *M. luteus* cells have a strict requirement for magnesium or manganese ions, the background "noise" of nonspecific DNA degradation could be largely, if not entirely, eliminated by simply including metal chelators in the reactions with UV-irradiated DNA. Furthermore, if one used heavily irradiated DNA as a substrate, the T4 and *M. luteus* enzymes degraded the DNA so extensively that much of it was rendered soluble after precipitation of the DNA with trichloroacetic acid. Since the simple measurement of the amount of acid-soluble radioactivity released from radiolabeled UV-irradiated DNA is technically very simple, this became a fairly standard assay procedure for the purification of these enzymes.

A series of events that began in 1974 led me to a second unintentional encounter with the base excision repair pathway. Sometime during that year, Ann Ganesan returned from a visit to Dick Setlow's laboratory, now located at the Brookhaven National Laboratory in Upton, New York. During this visit, Setlow informed her of his recent experimental evidence suggesting that the *M. luteus* enzyme was not dimer-specific after all. In his hands, partially purified preparations of the enzyme attacked DNA containing, of all things, uracil residues! Setlow and his longtime collaborator Bill Carrier had come across this curious observation during the course of experiments in which thymine in DNA was substituted by the heavy analog bromouracil, and some of the bromouracil had undergone debromination, yielding uracil. Even more convincingly, Setlow had resourcefully located a form of DNA in which thymine was naturally substituted by uracil—the genome of a bacteriophage called PBS2 which preyed on *Bacillus subtilis* as its natural host. When Ann Ganesan brought this news back to Stanford, I was flabbergasted. We had investigated many types of DNA base damage with highly purified T4 enzyme and had never observed a hint of endonuclease activity on any substrate other than DNA containing pyrimidine dimers. I was convinced that Setlow's partially purified preparations of the *M. luteus* enzyme were contaminated with something else. So, we prepared radiolabeled phage PBS2 DNA and to our pleasure found no degradation of this substrate by highly purified T4 UV endonuclease. But when we tested unfractionated extracts of *E. coli* (or *B. subtilis* for that matter), we observed extensive degradation of the uracil-containing PBS2 DNA. We concluded that Setlow's *M. luteus* preparations were indeed contaminated with an apparently ubiquitous endonuclease that degraded uracil-containing DNA and proceeded to explore two obvious questions. What is the precise nature of this new endonuclease, and more interestingly (at least to us), how on earth does phage PBS2 manage to survive and replicate its genome in *B. subtilis* in the presence of this potent degradative endonuclease?

The answer to the second question turned out to be surprisingly straightforward. When phage PBS2 infects *B. subtilis*, one of the earliest phage genes expressed encodes a specific and heat-stable inhibitor of the host enzyme that degrades uracil-containing DNA. This inhibitor has since been extensively purified and characterized by Dale Mosbaugh and his colleagues at the University of Oregon, and at the time of writing, this work has matured to the elucidation of the crystal structure of the inhibitor. Meanwhile, the answer to the first question emerged initially from Tomas Lindahl's laboratory in the Department of Chemistry at the Karolinska Institute in Stockholm. Lindahl's elucidation of the precise mechanism of action of the uracil-specific enzyme led to his recognition of an entirely new class of repair enzymes, now called DNA glycosylases, and to his discovery of base excision repair.

I first met Tomas Lindahl in the summer of 1971 at a meeting entitled *Molecular and Cellular Repair Processes*, which was organized by Roger Herriott and others at Johns Hopkins University. Having acquired an interest in enzymes that attack damaged DNA (the topic of Lindahl's presentation at that meeting), I took the liberty of introducing myself to him immediately following his talk. His response was typically gracious and engaging, and thus began a firm friendship that endures to this day. Our association has of course been largely motivated by our mutual interest in the mysteries of DNA repair, in particular our common enchantment with the intricacies of excision repair. But perhaps more memorable (to us) is our shared predilection for all manner of gastronomic delights, especially fine wines. As some of our colleagues will no doubt recall, particularly those who have joined us on such excursions, Lindahl and I have rarely passed up the opportunity to sample the culinary offerings of excellent restaurants placed intentionally or unintentionally at our doorstep in many parts of the world by the organizers of international conferences on DNA repair!

Tomas acquired a taste for fine wines during his days as a medical student in his native Sweden, and has since solicitously cultivated this taste to a refined art. During his tenure as a professor at the University of Gothenburg in the late nineteen-seventies, he was a wine taster for Scandinavia's largest popular magazine on wine and food. In Sweden, alcohol, including wines, can be purchased (legally) only in state-controlled stores. Every few weeks, the magazine issued telephone tasting reports of new wines received at these outlets through the medium of their wine tasters, of whom Tomas was the principal one. These chatty reports not only included reviews of wines, but also recommendations about their compatibility with specific foods, such as the delicious shrimp bountiful at certain times of the year off the Swedish coast. Although not directly paid for these services, Tomas was reimbursed for the purchase of as many fine wines as he wished (within reason) for comparative taste analyses. When I visited him at his home in Gothenburg during a sabbatical leave in his laboratory in 1978, I was astounded and unmistakably gratified by the munificence of his wine cellar. I can instantly evoke the flavor and texture of a particular richly gold sauternes enjoyed after a dinner with him and his former wife.

Lindahl received his early scientific training at the University of Stockholm Medical School, better known as the Karolinska Institute. While engaged in his medical studies, he joined a small research group in the Department of Microbiology. He was strongly influenced by his association with Einar Hammarsten who had just retired as Professor of Biochemistry. Lindahl described Hammarsten as "a small wiry man with intense gray-green feline eyes, and a total dedication to science." He recalled that on one occasion Hammarsten took him into the cold room to check on a chromatography column and began a lengthy scientific discourse, (apparently) totally oblivious of the fact that they were both freezing and that their discussion could just as readily have been conducted at ambient temperature.

During a postdoctoral stint with Jacques Fresco at Princeton University in the mid nineteen-sixties, Lindahl undertook a study of the conformation of a particular tRNA. While measuring the unfolding of this molecule at high temperature, his experiments were plagued by decomposition of the RNA that Lindahl eventually demonstrated to result from spontaneous degradation. He recalled:

> It occurred to me then that since DNA is a very much larger macromolecule which is present as a unique copy in bacteria, spontaneous degradation of DNA could be of physiological relevance. So I decided to investigate this point when I returned to the Karolinska Institute as assistant professor in Peter Reichard's department.

In the course of our correspondence, Lindahl pointed out that aside from its chemical and topological relationship to its complementary chain, a single DNA chain is itself a complex and intriguing macromolecule, the chemistry of which has largely been ignored from a historical point of view. He related:

> Whereas most scientists are aware of the important early contributions of Avery and Chargaff, fewer know that the key demonstration of a DNA chain as an unbranched polymer of nucleotides united in 3'-5' phosphodiester bonds was published by Dan Brown and Lord Alexander Todd in 1952. Without that crucial information, no reasonable model of the structure of the DNA duplex could have been built.

Lindahl related an amusing anecdote concerning this chapter in the history of DNA:

> In 1993, I had the pleasure of attending Dan Brown's 70th birthday party in Cambridge. The guest of honor, besides Dan himself, was the 86-year-old Lord Todd, who was awarded the Nobel Prize in chemistry in 1957. During a group conversation, a misguided molecular biologist chimed in, "Isn't it wonderful that this year is the 40-year jubilee of the structure of DNA!" Lord Todd magistrally corrected him in his strong Scottish accent: "Dan Brown and I discovered the *structure* of DNA. Crick and Watson worked on the *conformation*. But we didn't realize its biological implications."

Lindahl astutely recognized that an important minority of the many covalent bonds in a DNA strand are much less stable than the rest. By determining the rates of hydrolysis of those bonds in radioactively-labeled DNA as functions of temperature and pH, it became clear to him that spontaneous hydrolysis of DNA almost certainly occurs at a biologically significant rate under physiological conditions. This led him to the necessary conclusion that DNA repair enzymes must have evolved to counteract such spontaneous damage. For example, the deamination of cytosine to uracil would generate U·G mispairs in DNA which would be expected to yield G·C→A·T transition mutations during subsequent semiconservative synthesis, unless the uracil was somehow removed from the DNA. The deamination of cytosine seemed to him an especially interesting example of spontaneous base damage to pursue. As early as 1974, Lindahl commented in a paper that:

> Slow hydrolytic degradation of the primary structure of DNA occurs in neutral aqueous solution, and it seems likely that several types of spontaneous lesions are introduced into DNA at biologically significant rates under in vivo conditions. . . . The lability of cytosine in comparison with the other DNA bases raises the possibility that cells possess repair mechanisms to convert guanine·uracil base-pairs in DNA back to guanine·cytosine pairs.

So, Lindahl went in search of such DNA repair enzymes and in the process he discovered a novel biochemical mechanism by which the inappropriate base uracil is excised from DNA. The crucial element of his experiments was that, instead of simply measuring the formation of radiolabeled, acid-soluble products from a substrate containing radiolabeled uracil as Carrier and Setlow, and later I, had done, Lindahl characterized this fraction by chromatographic techniques and observed *free* uracil as a product of the reaction of substrate DNA with extracts of *E. coli*. Lindahl recounted that:

> I was lucky in having chosen, more or less accidentally, a chromatographic system that clearly separated uracil and deoxyuridine in product analysis. They often cochromatograph and this could have been misleading. I did not dare believe in the existence of such a novel enzyme until I succeeded in doing the technically more difficult experiment of showing that treatment of a polynucleotide substrate containing scattered U residues with a partly purified enzyme caused the introduction of alkali-labile sites but no chain breaks.

The enzymatic activity that catalyzed this reaction was the same as the one that we and Setlow and Carrier had detected earlier. But as Lindahl politely pointed out in a later review of the topic:

> Since most of these studies were concerned with other problems, the reaction products were not adequately characterized, and it was assumed that the degradation was due to cleavage of phosphodiester bonds adjacent to dUMP residues by an endonuclease.

The "studies" he referred to included one by Huber Warner and his colleague Merle Wovcha at the University of Minnesota. Warner was pursuing the solution to the potential problems caused by another mechanism by which uracil can arise in DNA—its incorporation by a DNA polymerase during DNA replication. As early as 1958, Arthur Kornberg and his colleagues had shown that the DNA polymerase he discovered in *E. coli* could efficiently utilize dUTP as a triphosphate precursor for DNA synthesis in vitro. Warner carried out DNA synthesis in vitro with "purified" *E. coli* DNA polymerase I in the presence of dUTP and observed nicking of the DNA at sites of uracil incorporation. He erroneously concluded that the DNA polymerase had an intrinsic uracil-specific endonuclease. Once again, uracil-DNA glycosylase had contaminated an enzyme preparation, and in the process another scientist's thinking! Arthur Kornberg is credited with the famous aphorism often quoted to me by David Goldthwait in my postdoctoral years, "Don't waste clean thoughts on dirty enzymes!" You will presently see how poorly this advice was heeded.

The enzyme-catalyzed release of free uracil from DNA requires the hydrolysis of the *N*-glycosyl bond linking the base to the sugar-phosphate backbone. Lindahl therefore named his new enzyme uracil *N*-glycosidase. Subsequent nomenclatural refinement enjoined a change to its present name, uracil-DNA glycosylase. Lindahl reflected that:

> The enzyme clearly acted in a hydrolytic way, hence the term "-ase." Additionally, the standard text books referred to the sugar-base bond as an *N*-glycosidic bond. But I received a couple of agitated letters from self-appointed nomenclatural experts who stated that the term "*N*-glycosidic bond" did not exist and the correct

term was "glycosyl bond." Initially I didn't pay much attention to these critics, but with increasing hassle about this (somewhat peripheral) issue I finally reluctantly agreed to change the term "glycosidase" to "glycosylase."

A short time after this discovery, my colleagues and I at Stanford showed that the enzyme in B. subtilis that attacked phage PBS2 DNA was identical in its mechanism of action and general properties. Since PBS2 DNA normally contains U·A base pairs, it was evident that the specificity of uracil-DNA glycosylase was not dependent on the presence of the U·G mispairs that result from the deamination of cytosine in normal C·G base pairs. Indeed, Lindahl later showed that the enzyme purified from E. coli even removes uracil from single-stranded DNA.

The loss of uracil (or for that matter any base) from double-stranded DNA poses a new problem for the cell, since the DNA now contains sites of base loss; so-called *a*purinic or *a*pyrimidinic (AP) sites. It was evident to Lindahl as early as 1974 that the complete repair of uracil in DNA must somehow address the coincident repair of the apyrimidinic sites generated by loss of that base. He was aware of the existence of an endonuclease in E. coli (then called endonuclease II), the independent discovery of which by two different laboratories will be recounted shortly. Endonuclease II was known to specifically attack phosphodiester bonds adjacent to sites of base loss and thus behaved as a prototypic AP endonuclease. Armed with this information, Lindahl suggested a formal biochemical pathway for the excision repair of uracil in DNA. In the paper published in 1974 in which he described the purification and characterization of uracil-DNA glycosylase from E. coli, he commented that:

> One possible function of this enzyme, acting in concert with endonuclease II, an exonuclease, a DNA polymerase, and DNA ligase, would be the reversion of guanine·uracil base-pairs in DNA to guanine·cytosine pairs by excision-repair.

Subsequent studies in several laboratories, including his own, demonstrated that the essential features of this pathway, predicated on the known catalytic functions of the five enzymes that Lindahl enumerated, were correct. Uracil in DNA is excised as the free base. The apyrimidinic site so generated is attacked by an AP endonuclease, creating a strand break with a sugar residue at one end. This residue is excised by an exonuclease and the resulting single nucleotide gap is restored by a DNA polymerase which incorporates the correct base cytosine, using the opposite normal DNA strand as an informational template. DNA ligase then restores the covalent integrity of the DNA. In 1976, my colleagues James Duncan, Lenore Hamilton, and I were motivated to suggest that this type of excision repair be called base excision repair (since it involves the excision of free bases), to distinguish it from the excision repair mode discovered by Setlow and Carrier years earlier, in which pyrimidine dimers are excised as oligonucleotide fragments. We suggested that the latter repair mode be called nucleotide excision repair.

Both DNA glycosylases and AP endonucleases are classes of enzymes that are unique to base excision repair of DNA. Once he discovered the former class, Lindahl sagaciously reasoned that the two enzymes might act sequentially. As already indicated, the discovery of AP endonucleases predated that of DNA glycosylases. The history of this discovery is a tortuous one which (regrettably) also touched my early scientific career in a less than spectacular manner. Several years prior to my tangential brushes with base excision repair through my observations of the enzymatic degradation of UV-irradiated DNA by the phage T4 UV endonuclease and the degradation of uracil-containing DNA, fate offered me a different entrée to this new mode of DNA repair during my tenure as a postdoctoral fellow with David Goldthwait at Case Western Reserve University in the late nineteen-sixties.

The history of scientific discovery is replete with examples of missed opportunity. In *The Eighth Day of Creation*, the noted science historian Horace Judson recounted a discussion with Francis Crick about the nature of this phenomenon. He suggested to Crick that "discovery, examined closely . . . seemed curiously difficult to pin to a moment or to an insight or even to a single person." Crick responded as follows:

> I think that's the nature of discoveries, many times: that the reason they're difficult to make is that you've got to take a series of steps, three or four steps, which if you don't make them you won't get there, and if you go wrong in any one of them you won't get there. It isn't a matter of one jump—that would be easy. You've got to make several successive jumps. And usually the pennies drop one after another until it all *clicks*. Otherwise it would be too easy!

Judson commented on the importance of what might euphemistically be called "the prepared mind." He quoted Sir Arthur Eddington's statement of 1934 that it is "a good rule not to put too much confidence in the observational results that are put forward until they are confirmed by theory," and reminds us that this "paradoxical inversion of the inductive system as preached from Bacon to Russell, has become an epigraph for the latter-day recension of the scientific method as practiced." An historically dramatic example of "why theory is indispensable to confirm fact" and how its absence can lead to missed opportunity is provided by Judson in recounting Erwin Chargaff's discovery of the base ratios in DNA, a discovery that Watson and Crick later acknowledged was critical to their elucidation of its structure. Chargaff's observations, based on measurements of the base composition of DNA from many organisms and species, was that while the absolute amount of the four bases A, T, G, and C varies from one DNA to another, the ratio of A:T and of G:C is always 1:1. As he put it in 1950:

> It is, however, noteworthy—whether this is more than accidental, cannot yet be said—that in all deoxypentose nucleic acids examined thus far the molar (that is, molecule-to-molecule) ratios of total purines to total pyrimidines, and also of adenine to thymine and of guanine to cytosine, were not far from one.

Judson remarked that:

> It is not easy to see how, at the time, Chargaff could have understood the significance of the equivalence rule or taken it any further; but it remains that he did not take it any further. So the observation he published and abandoned still smolders in his own recollection.

The opportunity that I missed as a young postdoctoral fellow was hardly comparable in importance to the discovery of the structure of DNA, so I wouldn't categorize my observations in Goldthwait's laboratory as "smoldering in my recollections" at this time. Nonetheless, the question lingers for me too. Having discovered an enzyme activity that degraded alkylated DNA, should we not have inferred the essential elements of the base excision repair pathway?

When I arrived in Goldthwait's laboratory as a first-year postdoctoral fellow in the winter of 1968, I was as raw as any fellow ever admitted to the Department of Biochemistry at Case Western Reserve University. This department was highly celebrated for the fundamental contributions of Harland Wood, Merton Utter, Warwick Sakami, and others to the biochemistry of intermediary metabolism, and counted among its many distinguished graduates scientists of the quality of Paul Berg and Jerard Hurwitz. Being a physician himself, Goldthwait was sympathetic to the notion that what I lacked in formal training in biochemistry and molecular biology might perhaps be offset by interest, enthusiasm, and sheer hard work—although I believe that there were moments during my first months in his laboratory when he considered such sympathies totally misguided. Goldthwait had established several areas of investigative activity in his laboratory. Having recently returned from a sabbatical leave with François Jacob in Paris, he was very excited about aspects of the lysogenic induction of the bacteriophage λ. While in Jacob's laboratory, Goldthwait worked on a mutant of *E. coli* that was thermosensitive for λ induction. This mutant was destined to play a significant role in the elucidation of the SOS response to UV radiation, the topic of Chapter 6. Goldthwait was also heavily involved in elucidating the biochemistry of RNA synthesis by *E. coli* RNA polymerase. Both of these projects were well under way when I arrived. He challenged me with something different. How did DNA molecules undergo recombination?

As already mentioned in reference to Bernie Strauss's interest in the topic, in the nineteen-sixties nothing was known about the biochemistry of recombination and this informational vacuum challenged the imagination of a number of investigators interested in DNA metabolism. A prevailing hypothesis at the time was that the process of homologous pairing between two DNA duplexes might involve specific localized distortions of the secondary structure of each duplex which somehow facilitated interaction between them. Goldthwait reasoned that if one could somehow simulate such distortions one might be able to observe strand exchanges in vitro. To my knowledge, Goldthwait was unaware that Strauss's forays into DNA alkylation had begun with his own curiosity about recombination. Nonetheless, Bernie's observation that extracts of *B. subtilis* or *M. lysodeikticus* attacked alkylated DNA were documented in the literature and this work tweaked the notion in Goldthwait's mind that alkylation damage in DNA might mimic the sort of "localized distortions" that may be productive for homologous pairing and hence recombination in vitro. After doggedly pursuing Strauss at a meeting to acquire information as to how one properly alkylated DNA, Goldthwait finally cornered him in a men's room, a position from which Strauss was apparently unable to retreat gracefully!

Being in no position to challenge the credibility of this hypothesis based on either knowledge or experience, I learned how to differentially radiolabel DNA molecules so that recombination intermediates could be detected by sedimentation in sucrose gradients, and set about to see what happens when alkylated DNA was incubated with extracts of *E. coli*. I observed no evidence for the much sought after "joint" molecules—putative recombination intermediates. However, I did observe significant degradation of the alkylated DNA, evidenced by a significant shift in its sedimentation profile to low-molecular-weight forms, as described earlier by Strauss. This degradation could also be observed by measuring acid-soluble radioactivity. This was the first occasion on which I made a crucial interpretative error; one that I was destined to repeat during my experiments with the phage T4 UV endonuclease several years later at Walter Reed, and again in my early years at Stanford with the enzyme activity that degrades DNA containing uracil. The error was common enough and was, I suppose, made in many laboratories at that time. I summed it up in the textbook *DNA Repair* published in 1985:

> For many years the release of acid-soluble radioactivity from radiolabeled DNA during its incubation with extracts was an indication of DNA degradation exclusively by endonucleases and/or exonucleases. However, the discovery of DNA glycosylases led to the realization that the formation of acid-soluble radioactive products can also result from the action of these enzymes. Hence, the simple measurement of total acid-soluble radioactivity does not distinguish between the release of free bases and the release of nucleotides from degraded DNA. A second potential source of confusion between DNA glycosylase and nuclease action on DNA stems from the historically well-established use of sedimentation of DNA in alkaline sucrose gradients for detection of the strand breaks produced by nucleases. However, the AP sites in DNA created by the action of DNA glycosylases are readily subject to β-elimination in the presence of strong alkali. Thus, sites of base loss are converted into strand breaks during the sedimentation, resulting in the potential for mistaken evidence of endonuclease action on the DNA.

Years later the wisdom of hindsight led to the recognition that during the reactions with *E. coli* extracts the alkylated DNA was attacked by a *series* of enzymes. One of these was another DNA glycosylase, "formally" discovered by Tomas Lindahl together with his postdoctoral fellow Riaz-ud-din, and independently by Jacques Laval and his colleagues in Paris, in 1977. This DNA glycosylase specifically excises free adenine residues alkylated in the N3 position. The resulting AP sites are then attacked by a potent AP endonuclease, rendering the DNA acid-soluble. Goldthwait and I purified and characterized the endonucleolytic component of this reaction which we designated endonuclease II of *E. coli*. (At the time, only one other DNA endonuclease had been discovered in this bacterium.) But, focused as we were on the process of recombination, we failed to recognize that this degradation resulted from the action of two enzymes acting sequentially—a DNA glycosylase specific for alkylated DNA and an endonuclease specific for the resulting AP sites. A further confounding observation that Goldthwait and I made at that time was that purified endonuclease II also attacked nonalkylated DNA to a limited extent. We naively considered nonalkylated DNA to be completely native in its structure, not realizing that our enzyme was in fact a powerful probe for the very small number of apurinic sites that spontaneously arise in DNA stored for any extended period. This observation not only further diverted our attention from the possible role of endonuclease II in the repair of alkylation damage, but in fact reinforced our notion of its possible role in recombination.

Shortly after I left Goldthwait's laboratory, a new postdoctoral fellow, Sheik Hadi, clearly established that in addition to alkylated DNA, endonuclease II degraded DNA containing sites of base loss. Hence, the degradation of "native" DNA was rationalized. Meanwhile, in the early nineteen-seventies, the Belgian biochemist Walter Verly, then at the University of Montreal in Canada, identified an AP endonuclease from *E. coli* exclusively based on its ability to degrade apurinic DNA. In subsequent studies, Verly extended these observations to other bacteria as well as to plants and animal cells. As recounted to me by Tomas Lindahl, Verly knew from his own work and from earlier studies by the English chemist Philip Lawley, an authority on DNA alkylation damage, that phage T7 exhibited what Lawley referred to as "delayed toxicity" after treatment with alkylating agents. Verly correctly deduced that this toxicity resulted from the slow spontaneous loss of alkylated purines from DNA at ambient temperatures, thereby generating sites of base loss, that is, AP sites. It occurred to Verly that the *E. coli* host might have an endonuclease activity that recognized such sites in the phage DNA and he searched for and found such an enzyme activity. In a paper published in the *Canadian Journal of Biochemistry* in 1972, Verly and his colleague Yves Paquette documented the existence of an endonuclease in *E. coli* that hydrolyzes apurinic sites in DNA. These authors also commented that:

> In further work to be shortly published, we have shown that the nuclease for apurinic sites of *E. coli* B41 has the same chromatographic behavior as endonuclease II, but our purified enzyme seems to have no action on alkylated sites. . . . *Escherichia coli* B41 thus contains an endonuclease that hydrolyzes at, or near, apurinic sites. . . . The endonuclease specific for apurinic sites could [also] play a role in the repair of depurinated DNA.

Progress in the nuclease field had been rapid and by this time the numerals III, IV, and V were usurped by other endonucleases discovered in *E. coli* extracts. So Verly called his enzyme activity endonuclease VI. Further studies in Goldthwait's and Verly's laboratories eventually showed that endonucleases II and VI were, in fact, the same enzyme. It is cogent to point out that Lindahl's model for what is now called base excision repair was not entirely dependent on the discovery of AP endonucleases by others. At about the same time that Verly and Paquette made their early observations on AP endonucleases, Lindahl made a similar independent observation. In a paper published in 1972 on the spontaneous cleavage of DNA strands at AP sites, he stated:

> Though it is concluded from this work that chain cleavage at apurinic sites in DNA seems to be a slow reaction under "physiological" conditions, cells may contain a special factor(s) that accelerates the rate of chain cleavage. . . . The most active factor in a calf thymus extract is a heat-labile endonuclease of mol. wt. ~40,000 that specifically attacks DNA at these lesions.

At the time that he wrote this paper, Lindahl encountered Verly's preliminary accounts of such an enzyme activity and cited them in his own manuscript. When Verly was preparing a definitive description of this class of enzymes (which he published in *Nature New Biology* a year later), he invited Lindahl to submit a companion report. Lindahl declined this invitation believing that their work would have greater impact if his group had a highly purified enzyme and presented the results as a more comprehensive study. So by Lindahl's admission "the excitement of having the priority of a conclusive report of a novel AP endonuclease rightly ended up with Verly."

Endonuclease II/VI had a history older than anyone suspected. Genetic evidence provided by Bernard Weiss and his colleagues, then at Johns Hopkins University, complemented by biochemical evidence from the Goldthwait laboratory, eventually showed that the AP endonuclease activity of endonuclease II/VI was a newly discovered catalytic function of a previously known enzyme called exonuclease III, discovered by Charles Richardson and Arthur Kornberg as early as 1964. Hence, much to the confusion of present-day students of DNA repair (including many of their professors), this AP endonuclease activity, which is the quantitatively major AP endonuclease of *E. coli*, is now called *exo*nuclease III after its accessory activity. Mercifully, the designations endonuclease II and endonuclease VI are no longer used in the literature.

Lindahl's discovery that a ubiquitously distributed enzyme that degrades DNA containing uracil is a DNA glycosylase cleared the way for the discovery of more enzymes of this class. We now know that different DNA glycosylases recognize multiple types of base damage in DNA, including that produced by the interaction of alkylating agents with the nitrogenous bases. This is the explanation for Bernie Strauss's original observation in the early nineteen-sixties, recounted earlier in this chapter, of the degradation of alkylated DNA by extracts of *M. lysodeikticus* and *B. subtilis*. Most surprisingly perhaps, even the phage T4-encoded UV endonuclease and that present in *M. luteus*, the enzymes that were so often mistakenly thought to represent the mechanistic homologs of the *E. coli* excision repair enzyme, were ultimately shown to fall into this class of DNA repair enzymes, a discovery that came about by a totally unexpected experimental observation.

In the late nineteen-seventies, Larry Grossman at Johns Hopkins and Bill Haseltine at the Dana Farber Cancer Center, together with their respective colleagues, elegantly exploited the use of DNA sequencing gels to validate the then prevailing hypothesis that the UV endonuclease purified from *M. luteus* catalyzed the hydrolysis of phosphodiester bonds immediately 5' to pyrimidine dimers, thereby setting the stage for excision of the dimers by exonuclease-mediated degradation of the DNA from the free ends generated by the endonuclease. Such incisions in a DNA fragment of known size and sequence containing randomly distributed pyrimidine dimers were expected to yield a ladder of DNA fragments by agarose gel electrophoresis, a technique recently popularized by the advent of DNA sequencing. The size of these fragments, and hence their electrophoretic mobility, could be precisely predicted based on the location of potential dimer sites that were attacked by the enzyme, determined by sequencing the DNA. To their initial confusion, the end-labeled fragments migrated in the gels as if they were approximately one nucleotide longer than anticipated. Grossman and Haseltine correctly reasoned that this electrophoretic pattern could result if, instead of cleaving the phosphodiester bond immediately 5' to a dimer, the *M. luteus* enzyme contained both a pyrimidine-dimer-specific DNA glycosylase activity that cleaved the 5' glycosyl bond within the dimerized pyrimidines and an AP endonuclease activity that subsequently cleaved the phosphodiester bond 3' to the resulting apyrimidinic site. This concerted DNA glycosylase/AP endonuclease-catalyzed attack would leave a sugar residue at the 3' end of each fragment, thereby accounting for the "anomalous" electrophoretic anomaly. The direct proof that only one of the two glycosyl bonds in the dimerized pyrimidine pair was cleaved was achieved by showing that free thymine was released if the incised DNA was treated with photoreactivating enzyme,

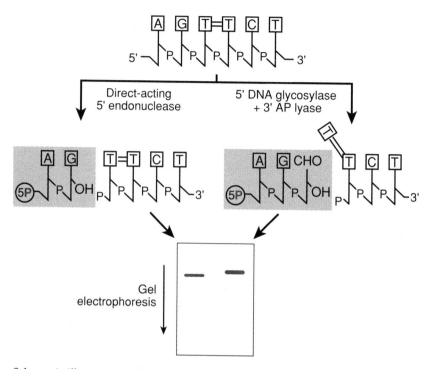

Schematic illustration of how the *M. luteus* pyrimidine dimer DNA glycosylase activity was discovered. For simplicity, only one DNA strand is shown. If DNA radiolabeled at its 5′ end (*shaded area*) and containing a pyrimidine dimer were cleaved 5′ to the dimer by a 5′ endonuclease (*left*), DNA fragments of a particular size distribution are expected in the sequencing gel. However, the fragments observed following incubation with the *M. luteus* enzyme migrated in sequencing gels as if they were approximately one nucleotide larger, suggesting the cleavage mechanism on the right. (Reproduced, with permission, from Friedberg et al. 1995, copyright by ASM Press, Washington, D.C.)

which monomerized the dimer. Eric Radany, a graduate student in my laboratory at Stanford, obtained the same results with the phage T4 UV endonuclease. I was relieved to finally be rid of the scourge of having stumbled onto three different DNA glycosylases and failing to immediately recognize any of them as such!

As the decade of the nineteen-seventies drew to a close, it was evident that nature exploits multiple mechanisms for the repair of base damage. It was particularly striking that the repair of pyrimidine dimers could be effected by three biochemically distinct pathways—photoreactivation, nucleotide excision repair, and base excision repair initiated by the phage T4 or *M. luteus* pyrimidine dimer-DNA glycosylases. UV radiation damage from the sun was clearly a potent selective agent during biological evolution. But there was further gold to be mined from nature's repertoire of DNA repair strategies. A particularly brilliant nugget emerged with the discovery of a novel DNA repair pathway that handles small alkyl groups located specifically at the O^6 position of the base guanine. The repair of this lesion turned out to require a most unusual enzyme—one that was completely different from any thus far encountered in the rapidly expanding DNA repair world. The beginnings of this story and the essential biology that laid the ground work for the later biochemistry involved another eminent figure in molecular biology, John Cairns.

Molecular biology owes a huge debt to medical schools and departments of physics around the world, which served as the initial training grounds for many of its illustrious sons and daughters. Like Goldthwait, Lindahl, and the Nobelists Salvador Luria, Renato Dulbecco, and Arthur Kornberg, John Cairns received his formal training as a physician. At an international symposium on DNA repair in Taos, New Mexico in 1995 at which Cairns was featured as the keynote speaker, he recounted that the first time he met Max Delbrück he was asked about his background. He told Delbrück that he was an M.D., at which point Delbrück responded with a disdainful "Oh!" and walked away. "Which made the point very nicely," rejoined Cairns. A man of eclectic scientific tastes and an extraordinary bent for new and different scientific horizons, Cairns has spent his professional career on the continents of Africa, Australia, North America, and Europe. "I look back," he told his Taos audience with a characteristic twinkle in his clear blue eyes, "and each of the times that I've gone from one continent to another to do some project my life has totally changed and was never the same again. Each of these was a quantum jump forward for me; or at least a quantum jump up in interest in my scientific life—and that is the important question—is your scientific life interesting?"

Cairns is perhaps best remembered for his several notable contributions to our knowledge of DNA replication, a field to which he had a memorable introduction in 1957 during a four-month minisabbatical at Caltech. He told the audience in Taos:

> I had the fortune to stay initially at the Faculty Club until I discovered that I couldn't pay the bill; four days had used up half my capital. So I fled to somewhere cheaper. It was a house occupied by some graduate students. The graduate students happened to be Jan Drake, Mat Meselson, Howard Temin, and until very recently Frank Stahl. The whole house was engaged desperately in discussing the meaning of an experiment going on at that time; the Meselson-Stahl density transfer experiment. I cannot imagine a nicer way of being introduced to molecular biology.

More about this famous experiment in the following chapter!

Several years after returning to Australia, where he planned "to give a two-hour seminar updating everyone on advances in molecular biology circa 1953–1957," Cairns became eligible for a full sabbatical, which he elected to take with Al Hershey at Cold Spring Harbor Laboratory. There was considerable uncertainty at the time about the structure of the unit DNA molecule and exactly how it replicated, and Hershey was attempting to devise methods for isolating intact DNA molecules. The first question that Cairns tackled was whether the intact phage T2 DNA molecules isolated by Hershey were indeed double-stranded molecules of the expected length. This was done by autoradiography of DNA, a technique that Cairns had developed in Australia in order to study the replication of vaccinia viral DNA. Cairns decided that it would be interesting to determine whether "you could see molecules caught in the act of replication" by labeling *E. coli* with radioactive thymine and preparing autoradiographs of cells that had "ever so gently spilled out their DNA." These technically difficult and tedious experiments with *E. coli* were initiated at Cold Spring Harbor and came to fruition when he returned to Australia. He recounted:

> I remember getting very depressed. These autoradiographic experiments are rather foul because you collect the DNA and put it out on the film and you have to leave it there for two months incubating away; collecting grains. No sooner had I put an experiment away to incubate than I would think that I should have done it a better way. So I would do another experiment, and lose interest in the previous one. It was a life of continual frustration and annoyance. I remember that I used to go to the coast in Australia for holidays and would walk constantly up and down the beaches trying to find out how often a nylon fishing line was washed up on the beach untangled—and never found one! But in the end I did find one or two molecules of *E. coli* DNA which appeared to be more or less untangled.

In a display of modesty that I suspect Cairns has cultivated with fierce pride over the years, he refrained from showing his Taos audience pictures of the now famous autoradiographs of *E. coli* circular chromosomes "caught in the act of replication."

The powers that be at Cold Spring Harbor were clearly impressed with Cairns. At the tender age of 40 he was invited to become the new director of the Laboratory, a task that consumed him for the next several years. "Fortunately, I had time to complete the autoradiographic studies on replicating *E. coli* chromosomes before leaving Australia for Cold Spring Harbor and its new massive responsibilities," Cairns later confided to me, "otherwise I would never have got them done." Colleagues in the field who are intimately knowledgeable about these tedious and time-consuming autoradiographic experiments might often have wondered how in a few short years Cairns could find the time amidst his administrative responsibilities additionally to define the existence of multiple short replicons in the DNA of human cells. What transpired is that after a few "horrifying" years as director of Cold Spring Harbor Laboratory, Cairns received a phone call from Matthew Meselson, who wanted to find out something about the structure of eukaryotic DNA. Cairns recalled:

> He proposed that I grow *Drosophila* larvae in tritiated thymidine. He would then purify one of their chromosomes and I would do the autoradiography. I told him that that would require huge amounts of tritium and be foully dangerous. Why not simply do the whole thing with HeLa cells? And as I said that, I suddenly remembered that I had done precisely that three years earlier, just before leaving Australia, and the developed slides were in the drawer by the phone, waiting to be looked at. "I'll call you back," I told Matt. Two weeks later, I sent off a letter to the *Journal of Molecular Biology* describing multiple short replicons in HeLa DNA!

In 1968, relief came to Cairns in the person of Jim Watson, who took over the directorship at Cold Spring Harbor and Cairns was "free from having to raise money." So he began his next big adventure with DNA replication, the isolation of a mutant of *E. coli* defective in the only known DNA polymerase at that time, which to everyone's surprise "lived to tell the tale." Cairns relates his interest in isolating such a mutant to his general misgivings about the reported rate of replication of the *E. coli* chromosome. He was not convinced that the Kornberg DNA polymerase (as *E. coli* DNA polymerase I was, and still is, often referred to) replicated DNA fast enough to account for the known rate of DNA replication in vivo, and he was also not persuaded that the number of DNA polymerase I molecules in *E. coli* could reasonably accommodate this deficiency by simple mass action. Cairns's growing conviction that perhaps the Kornberg polymerase was not the enzyme that carried out semiconservative DNA synthesis grew when he heard that Roy Curtis III at the Oak Ridge National Laboratory had isolated a mutant of *E. coli* that segregated small daughter cells (so-called minicells) that contained no DNA at all, but had normal levels of DNA polymerase. Cairns remarked:

> This said to me that the 400 molecules of DNA polymerase were not busy replicating DNA. They were just floating around. Instantly the thought came to me that maybe they were not there to replicate DNA, but were rather for casual repair and tidying things up. So I thought that we should be able to isolate a viable mutant that was missing DNA polymerase.

And indeed they did. Cairns spent six months devising a cunning screening assay, and his capable technical assistant Paula DeLucia spent another six months looking for the desired mutants and found one after screening a little over 3000 candidates. Cairns told us in Taos:

> We were very excited about this. Julian Gross had been teaching the bacterial genetics course at Cold Spring Harbor, and I told him that we ought to name it after Paula DeLucia in some way so that people will remember who did it. Gross said, "That's easy, just call it Paula." That's why the mutant strain was called *polA*!

The viability of this mutant not only disabused the scientific world of the notion that DNA polymerase I was the replicative polymerase in *E. coli*, but additionally opened the doors for the subsequent discovery of DNA polymerases II and III, the latter being the enzyme required for semiconservative DNA synthesis. "Incidentally," Cairns later wrote to me, "it turned out that Sohei Kondo [an eminent Japanese geneticist] had already isolated a *polA*-defective strain of *E. coli*, but didn't know it because he didn't test it for DNA polymerase I activity."

Following his stint at Cold Spring Harbor, Cairns crossed the Atlantic to his native England to assume the directorship of the Imperial Cancer Research Fund (ICRF) Laboratories at Mill Hill, London. Younger readers unfamiliar with the structure and organization of various research enterprises in the United Kingdom might be interested to know that the ICRF is one the most privileged environments in the world in which to pursue a research career. Financial support to the Fund (which is private and hence operates by its own rules) derives from charitable contributions which are used to support the investigative efforts of leading scientists with full-time (and often tenured) positions at various ICRF Laboratories scattered around Britain. ICRF scientists have no formal teaching responsibilities, although they can enjoy the labors of outstanding graduate students affiliated with Ph.D. programs at regional universities. Additionally, much to the envy of their university peers, ICRF scientists are not required to obtain research funds by the traditional manner of competitive grant application, although they are certainly not constrained from this endeavor. Hence, to be an ICRF investigator is the closest thing to scientific heaven in Britain, threatened only by the vicissitudes of the generosity of the British public, a threat hardly to be taken seriously until all known cancers can be cured.

With his characteristic flair for new horizons, Cairns focused his intellectual talents on the cancer problem, a commitment that he considered appropriate and necessary for the director of an institute dedicated to cancer research. In 1978, he published a small book entitled *Cancer—Science and Society*, intended mainly for the intelligent general reader. In the preface, he referred to this book as a highly abbreviated version of what he originally intended as "a compendious review of the whole subject—the work I wished I myself could have read on first encountering the field."

One of the many aspects of carcinogenesis that intrigued Cairns was the notion that the incidence of cancer in any particular species is not solely determined by its exposure to mutagens. In the Leeuwenhoek Lecture delivered to the Royal Society in London in 1978, he posed the argument as follows:

> For example, although the mutability of human and rodent cells *in vitro* is about the same, the chance that any particular cell becomes cancerous in any set interval of time is far higher in rodents than in humans. In terms of the individual steps in the process of carcinogenesis, the difference could amount to a factor of 10^9 or more. A small part of this colossal difference may be attributable to species differences in the efficiency of DNA repair and part may be related to the fact that human cells are, for some reason, much less ready than rodent cells to become transformed after exposure to mutagens *in vitro*. But it is obviously worth considering what other factors may be operating to prevent long-lived animals from accumulating somatic mutations or to protect them from the consequence of mutation.
>
> Some years ago I tried to draw up a list of the ways in which multicellular creatures might control and organize their cell lineage in order to minimize their rates of accumulation of mutations.... One conclusion that I came to was that the rate of accumulation of mutations in self-renewing epithelia could be diminished if there were strict rules governing the segregation of sister chromatids at cell division.

In 1974, Cairns encountered a student who had just completed her undergraduate training at the University of Aberdeen. Having somehow missed the application deadlines to graduate programs in most of the better universities in the United Kingdom, Leona Samson, in something of a panic, hurriedly applied to the ICRF and interviewed with two of the staff, one of whom was Cairns. Her apparent naiveté about the procedure for becoming a graduate student at a university extended to her ignorance of her future mentor's considerable scientific fame. At that time, Cairns was attracted to an experimental bacterial replication system devised by Charles Helmstetter and Donald Cummings in 1963, which he viewed as being formally analogous to a self-renewing mammalian tissue, such as the epithelium of the skin or intestine. In this experimental system, an asynchronous culture of bacteria was placed on a Millipore filter. The filter was inverted and medium was passed through the back. Bacteria that stuck to the filter remained there, but the rest were washed off. Eventually, only a single layer of cells "packed shoulder to shoulder," as Cairns described the system to me, was left on the filter. As these cells divided, half of the daughter cells remained bound to the filter while the other half were carried off into the medium. The system could thus be viewed as a bacterial model of the basal layer of dividing cells found in many mammalian cell epithelia.

> We thought it possible, therefore, that bacteria stuck on a filter might apportion their daughter chromosomes in a non-random fashion so that one particular tem-

plate strand always stayed associated with the filter through successive generations. So we felt that we might be able to use such a filter system both as a means for studying the segregation of mutations and as a model for the behavior of epithelia.

As is the fate of many elegant scientific hypotheses, this one was ruined by the immediate observation by Samson and Cairns that the bacteria segregated their chromosomes randomly at each cell division. Nonetheless, Cairns was sufficiently persuaded that the experimental system that he and Leona Samson had devised (which he termed a "topostat," but which, according to Samson, was affectionately referred to in laboratory jargon as the "baby machine" because it constantly released newly divided cells) had the potential for answering other important questions about mutagenesis. He therefore (with some difficulty) persuaded Leona Samson to set about devising a way of testing whether there were any particular rules to mutagenesis that could be manipulated by multicellular organisms to minimize the mutation rate in stem cells. In the published account of his lecture to the Royal Society, he wrote that:

> Such a "topostat" should not be subject to the phenomenon of periodic selection, in which variants with higher growth rate are continually supplanting the existing population; presumably a topostat (or any ordered epithelium if it comes to that) should be largely protected from the forces of natural selection. . . . [Another] question that we hoped to answer was whether mutation rates were simply proportional to mutagen concentration even at low concentration.

When I talked with Leona Samson, now a professor in the School of Public Health at Harvard and an eminent scientist in her own right, she remembered the issue of the relationship of mutation rates to mutagen dose as a particularly compelling leitmotif in Cairns's thinking. She recalled:

> He often talked of the fact that most carcinogenesis experiments used very large doses of carcinogens for short periods of time in order to get lots of mutations and tumors, whereas in the real world people were exposed chronically to low doses of mutagen. So he tried to convince me that I had the perfect system to measure how mutants accumulate in a growing population that continues to be exposed to low doses of a mutagen. He had to work hard to persuade me to do these experiments because I thought it would be very dull to just look at the rate of the accumulation of mutants at different doses of carcinogens.

Her reservations notwithstanding, Samson set up continuous exposure of *E. coli* stuck on filters to the chemical mutagen N-methyl-N'-nitro-N-nitrosoguanidine (MNNG), an agent known to alkylate DNA, and measured the mutation rate in the effluent cells as a function of time. Cairns told his Royal Society audience that:

> From the first the results were totally unexpected. When MNNG was added to the medium flowing through the filter, the frequency of mutants in the output of the filter did increase, but all the increase was confined to the first hour of exposure to mutagen. The failure of the mutagen to work after the first hour could not be due to any selection of mutagen-resistant variants because the filter continued, throughout the whole experiment, to produce a constant number of viable cells in each time interval. So all the bacteria had to have become either impermeable to the mutagen or resistant to it in some other way. It soon became likely that we were dealing with some form of DNA repair. . . . For the next three years, therefore, the laboratory

gave itself over to the study of this reaction, which we called the "adaptive" response.

In a typically self-effacing letter to me in 1982, Cairns wrote that "the rest was, in a manner of speaking, routine plain sailing. The obvious alternate explanations were excluded and she [Leona Samson] (and the others) were off to the races." But from the eager perspective of a young graduate student preoccupied with completing a Ph.D. thesis, all was not that "plain sailing." Leona related to me that she:

> Spent another six months doing these experiments over and over again because the filter culture refused to accumulate mutants. And of course I assumed that it was because I was doing something wrong. I remember the day that we sat down with all this data and said "We've got to try and make sense of what's going on here." I don't remember who specifically came up with the idea that the E. coli cells might become resistant to the mutagen. But I went off to check that. The first experiment gave a spectacular result!

The seminal observation of the now well-known adaptive response to alkylation damage was delineated by Samson and Cairns in 1977 in a paper entitled "A New Pathway for DNA Repair in *Escherichia coli.*" In a series of subsequent experiments, Cairns and Samson, together with Cairns's postdoctoral fellows Penelope Jeggo, Martine Defais, Paul Schendel, and a highly skilled technical assistant, Peter Robins, elaborated the essential features of this adaptive response. They showed that adaptation was effected by exposure of cells to low levels of alkylating agents and manifested as an increased resistance (adaptation) to killing and to mutagenesis by subsequent exposure to higher concentrations of the alkylating agents. In their initial experiments, Samson and Cairns showed that this "adaptive" phenomenon required protein synthesis, providing the first hint that it was an inducible response. The details of the complex regulation of this induction were later revealed in an elegant series of experiments in both Tomas Lindahl's and Mutsuo Sekiguchi's laboratories during the nineteen-eighties. This is a fascinating bit of recent history that will not be recounted here.

In his 1982 letter to me, Cairns volunteered the sobering reflection that:

> It might have been hard to make this discovery any other way. Short of monitoring the *rate* of increase of mutations in relation to constant MNNG concentration in continuing growing cultures (which is not easy to engineer for a mutagen as unstable as MNNG), any inducible repair will only be picked up if you correctly guess *both* the inducing concentration *and* the challenge concentration. Conceivably these could be rather close together.

It subsequently became established by the isolation of appropriate mutants that the genes that were induced in adapted cells encoded (at least) two types of DNA repair enzymes. Work in several laboratories showed that adaptation to killing results from the enhanced repair of alkylation damage by yet another DNA glycosylase, one that specifically recognizes certain alkylated bases and is encoded by a gene called *alkA*. However, the enzyme that provides resistance to the mutagenic effects of alkylation damage, encoded by a gene called *ada* (for *ada*ptive response), turned out to be novel in the DNA repair world. In the interests of sparing the uninitiated reader possible confusion, let me point out that the Ada protein is endowed with the ability to specifically recognize O^6-methylguanine and O^4-methylthymine in DNA. It plucks the methyl groups from these (and only these) two positions of these (and only these) two alkylated bases in DNA, transferring them to a specific amino acid residue in the polypeptide. This covalent methylation of Ada protein inactivates the enzyme, which therefore can only act once. Hence, Ada protein is an example of a class of enzymes called suicide proteins. So, protection (adaptation) from the mutagenic properties of certain alkylating agents is only as effective as the available supply of Ada protein in the cell. How did this amazing story unfold?

Very soon after the discovery of the adaptive phenomenon, the Cairns group learned of the earlier work of Tony Loveless and Philip Lawley in London suggesting that of the multiple alkylation products in DNA, O^6-methylguanine and O^4-methylthymine were the two that were most likely implicated in mutagenesis because these could most obviously interfere with the normal base pairing properties of G and T, respectively. Indeed, O^6-methylguanine and O^4-methylthymine had been directly shown to mispair during DNA replication and transcription in vitro. Paul Schendel and Peter Robins carefully monitored the amounts of various alkylated bases in the DNA of adapted and unadapted bacteria. They found that even though O^6-methylguanine was a quantitatively minor alkylation product in DNA, it underwent a striking quantitative reduction in adapted cells. A potential target for the adaptive response to mutagenesis was thus identified. However, most disturbingly, this "tolerance" to the accumulation of the alkylated base in DNA was limited. After exposure to higher levels of MNNG, even the adapted cells began to accumulate O^6-methylguanine. We now know that this reflects the saturation of all available Ada protein in the cell, but that was by no means obvious at the time. The Cairns group therefore embarked on a series of experiments to measure the detailed kinetics of the disappearance and reappearance of O^6-methylguanine in cells exposed to various levels of alkylation damage.

While Robins was busily executing these kinetic experiments at the Mill Hill Laboratories in London, Cairns, much to any historian's delight, systematically communicated his thoughts about adaptation in a series of letters to Paul Schendel, who had since moved to the University of California in Berkeley to pursue postdoctoral training. A letter written in late November 1978 is particularly illuminating about Cairns's insights concerning the nature of the protein encoded by the *ada* gene, which repairs O^6-methylguanine in DNA, a protein that he and his colleagues affectionately and abstractly initially dubbed "preventase." "Preventase was the word that we used when we thought that the Ada protein might actually protect guanine from O^6 alkylation," Cairns revealed. Leona Samson elaborated on this historical nuance:

> When we first found that exposure to low doses of alkylating agents caused cells to acquire resistance to agents like MNNG, our first priority was to figure out whether resistance was achieved by *preventing* DNA alkylation damage or by *repairing* alkylation damage. The two models became known in the lab as "preventase" versus "repairase." In our first paper, we excluded the preventase model because alkylated plasmids survived better in adapted cells than in nonadapted cells, so it was clear that adapted cells handled alkylated DNA better than nonadapted cells.

Here is an excerpt from the letter that Cairns wrote to Paul Schendel on November 23, 1978:

Dear Paul,

The story advances:

(1) Pete [Robins] now has the longer-time experiment and the results are roughly what I would have expected. . . . It seems to me, therefore, that the simplest conclusion is that adaptation lowers the initial (starting) level of O6s but does <u>not</u> markedly alter the subsequent rate of removal of O6s (at least when that is expressed as O6s removed per bacterium per second, which is surely the natural way to do the calculation). Many more experiments would have to be done to establish whether adaptation had <u>any</u> effect on the subsequent removal of O6s, and I don't think these experiments are worth doing at the moment.

Presumably the "preventase" effect is really some very rapid excision by something that is consumed in the process (i.e., is not simply an enzyme). I may have mentioned to you the thought that the active principle could be rather like an antibody to O6MeG within DNA. . . .

So the story evolves

John.

In later correspondence, Cairns confided to me that:

What really took a huge amount of thought was my deduction that the inducible repair of O6MeG was being done by an enzyme that acted only once. It took 6 months of contemplation of Paul Schendel's results showing that O^6-methylguanine was removed from the DNA of adapted cells, plus a lot of correspondence, followed by the appropriate experiments. What made it hard was that I did not know of any example of an enzyme that worked only once. After I had written the paper Rob Kay told me about once-off restriction enzymes and then later, suicide substrates were described.

The detailed elucidation of the nature of the suicide Ada protein and how it actually worked required a frontal assault by a biochemically-oriented laboratory. The protein had to be identified, purified, and characterized. This task was achieved by Tomas Lindahl and his colleagues, who had by this time identified a strong interest in the repair of alkylation damage in DNA based on their earlier studies on DNA glycosylases that specifically repair alkylated DNA. In his letter of November 23, 1978, Cairns informed Schendel that Lindahl's assault on the biochemical mechanism of adaptation to alkylation mutagenesis had begun:

Yesterday I was called up by Tom Lindahl who was in some state of excitement because he has got the adaptive removal of O6s to work in vitro. His experiments are really very pretty. He methylates DNA in vitro using hot MNUA. This DNA is then incubated overnight at 80° (I forget what pH). This removes the N7s, and so he is left predominantly with O6s in the DNA. The DNA is then mixed with extracts from unadapted and adapted bacteria.

If he then treats this DNA-plus-extract mixture with DNAse, he observes that there are no longer any O6-Me-GMPs present when the extract comes from adapteds. However, he was astonished to find that if, instead of treating with DNAse, he precipitated the mixture with ethanol, the liberated O6s were not in the supernatant. In fact, after the DNAse treatment they proved to be at the origin (in the TLC plates, or whatever he was using to separate the nucleotides).

At this point, he started to wonder if the adaptive enzyme were moving the methyl from the O6 position onto the phosphates and therefore preventing the DNAse from acting. However I told him that the essence of the adaptive mechanism is that it is consumed. So I said I thought he would find the missing methyl groups attached to some specific protein.

Letters to *Nature* from the Cairns and Lindahl laboratories describing the kinetics of the loss of O^6-methylguanine from the DNA of adapted *E. coli* cells and the attendant prediction of the stoichiometric nature of the repair enzyme, as well as the demonstration of the repair of O^6-methylguanine in vitro, were published back-to-back. In his paper with his postdoctoral fellow Peter Karran, Lindahl declined to speculate on Cairns's suggestion that the methyl group might be transferred to a specific protein. In recent correspondence, Lindahl told me that he had cultivated an interest in the repair of O^6-methylguanine dating back to Philip Lawley's work in 1970 showing that this alkylation product in DNA is rapidly removed from DNA; more rapidly than could be accounted for by spontaneous hydrolysis. However, Lindahl was deliberate in pointing out that "we were getting nowhere with the biochemistry of this problem until Samson and Cairns discovered that the repair activity is inducible." Cairns recalls that:

> We had worked out from the kinetics that a consumable factor likely accounted for the removal of methyl groups. I told Tomas this when he visited the neighboring MRC laboratory. He said that the repair would be due to an enzyme, and I said no, it is due to something that acts only once.

This debate is not really about the petty issue of priority for a discovery, but more interestingly (to me) reflects the alternative philosophic views of the relative "importance" that some scientists attribute to in vivo (genetic) versus in vitro (biochemical) approaches to the elucidation of biological functions. After reading a draft of this chapter, Cairns, whose opinions I respect highly, commented that: "I found your chapter much more biochemical and less biological-genetical than I would have made it. After all, the only certain way one has for determining which enzymes are responsible for any particular form of repair is to isolate mutants that are defective in one or other step in the process—and this is something you barely mention." In defense of biochemistry, for which I have an admitted bias, I responded to him that: "I can give your criticism a different twist by offering the equally valid statement that the only certain way one has for knowing how a gene that is responsible for a particular form of repair operates, is to isolate the protein that it encodes and characterize it!" Hopefully student readers who are beginning their careers as molecular biologists will recognize that both arguments are biased. Biochemistry in the absence of a genetic framework is essentially meaningless, and genetic predictions without biochemical validation are, in the final analysis, just hypotheses.

Lindahl related:

> The resolution of the unique mechanism by which the Ada protein works was challenging. We already knew that O^6-methylguanine was not repaired by a DNA glycosylase, having looked carefully for such an activity in parallel with our studies on the repair of 3-methyladenine, which we knew was excised by a DNA glycosylase. Moreover, our finding that in cell extracts of adapted *E. coli* O^6-methylguanine disappeared from DNA without the release of detectable amounts of free base immediately established that a DNA glycosylase was not involved. We made at least one serious false step in the beginning. We could show that O^6-methylguanine unexpectedly disappeared from DNA without any release of

methanol, free base, or nucleotides. However, in trying to define the acceptor site we were initially absorbed with the erroneous idea that the methyl group was moved to another, less harmful site in DNA.

Lindahl and his colleagues had in fact considered the possibility that the methyl groups were transferred to a protein and tested this idea. He recounted:

We treated our reaction mixture containing large amounts of crude extract of adapted *E. coli* with pronase at the end of the reaction, but saw little or no release of radioactive material. But we failed to realize that the active protein acceptor domain is fairly protease-resistant and that we had not obtained complete proteolysis. [A year later] we corrected our mistake because once we purified the activity, it became clear that the methyl group was removed from DNA to a protein, as suggested by John Cairns.

Lindahl and his assistant Monica Olsson used radioactively tagged methyl groups to show that the "specific protein" referred to by Cairns was the same as that which catalyzed the removal of the methyl groups from the O^6 position of guanine in the first place. This enzyme (now called O^6-methylguanine-DNA methyltransferase) shifts the methyl group to a particular cysteine residue in the protein, thereby inactivating itself. Cairns was correct—the enzyme only worked once and then committed suicide by methylating itself.

The definition of the particular methyl acceptor residue in the transferase protein was another tricky problem. Lindahl informed me that:

Four amino acids were known to be acceptors of methyl groups in various biological transmethylation reactions; lysine, arginine, histidine, and glutamic acid. So we assumed the relevant amino acid should be one of these four. The first three could be ruled out by chromatography and amino acid analysis, but the methylated residue from the transferase and methylglutamic acid practically cochromatographed. Thanks to the considerable technical skill of Monica Olsson, we were not entirely satisfied with having identified methylglutamic acid and fortunately did not publish these results. Instead, we continued searching for previously unknown amino acid acceptors and finally identified S-methylcysteine as the relevant residue in the protein.

The adaptive response to alkylation damage has turned out to be one of the most intriguing examples of DNA repair in *E. coli*. The O^6-methylguanine-DNA methyltransferase is not only a repair enzyme, but additionally when methylated at a different cysteine residue it assumes the role of a positive regulator of the gene that encodes the protein, and of other genes involved in the adaptive response. Hence, when cells are exposed to alkylating agents, this regulatory response ensures the rapid expression of increased amounts of multiple gene products which collectively effect the repair of different types of alkylation damage in DNA. Although clearly mechanistically quite distinct from the splitting of pyrimidine dimers by photoreactivating enzyme, the repair of O^6-methyl guanine and the monomerization of dimers both represent examples of DNA repair by the direct reversal of base damage, thereby circumventing the need to excise and replace damaged bases and avoiding any potential for introducing errors in the DNA sequence.

CHAPTER 5
Mismatched Bases Are Repaired by Yet a Different Mode

MISMATCH REPAIR IS A PROCESS whereby mispaired bases in DNA are restored to the correct nucleotide sequence. Mispaired bases most frequently arise as a consequence of errors during the process of DNA replication, and if not corrected they typically result in permanent alterations in genetic information. For example, imagine that during the synthesis of a new DNA strand, the replicative machinery stably incorporates the incorrect base cytosine (C) instead of the correct base adenine (A) opposite the base thymine (T) in the parental DNA template. When the double-stranded DNA molecule carrying this C·T mispair is subsequently replicated, the base C will, of course, correctly instruct the formation of a C·G base pair, thereby effecting a permanent change from an T·A to a C·G base pair and hence a permanent alteration in the genetic information carried by those DNA molecules. The repair of mismatched bases before they imprint the genome with alterations in nucleotide sequence is a singularly important mutational avoidance mechanism in living cells. As is often the case in the evolution of scientific disciplines, the investigative impetus that led to the discovery of mismatch repair as a mutational avoidance mechanism during DNA replication was indirect. It did not emerge from a primary interest in mutagenesis per se as much as from a curiosity about the mechanism of genetic recombination and the imperative to explain certain aberrations of this process, such as gene conversion, a term that I will define more precisely shortly.

The tortuous path from recombination to mismatch repair was pursued independently by several teams of investigators who were studying aspects of the exchange between DNA molecules in different contexts and in different experimental systems. Let's begin with Sanford (Sandy) Lacks, who in 1955 as a graduate student joined the laboratory of Rollin D. Hotchkiss at the Rockefeller Institute in New York. A native of New York City, Lacks began his professional education as an engineering student, but decided that he wanted to go to medical school, so he transferred to a premedical program at Union College in Schenectady, New York. Here, his exposure to biochemistry and genetics provoked an increasing interest in basic biological research. At about this time, Detlev Bronk, then president of the Rockefeller Institute, was conducting a vigorous recruiting campaign at several colleges in the northeast to attract outstanding graduate students to the Institute. The president of Union College recommended Lacks enthusiastically as a candidate for graduate training at the Rockefeller, and he was offered and accepted a position there. Lacks recounted:

> When I arrived at Rockefeller, I heard an orientation lecture by Rollin Hotchkiss that deeply impressed me. Also, everyone I spoke to who had worked with him held Hotchkiss in very high regard. So I decided to join his laboratory.

Rollin Hotchkiss, as well as Harriet Taylor, whom we will presently encounter as the more familiar Harriet Ephrussi-Taylor (she married the French geneticist Boris Ephrussi), were both trained in Oswald Avery's laboratory at the Rockefeller. You will recall from Chapter 1 that Avery, together with Maclyn McCarty and Colin MacLeod, showed that the so-called transforming principle that could permanently alter the genetic constitution of certain strains of pneumococcus was, in fact, DNA. After the initial experiments in Avery's laboratory on the transformation of a nonencapsulated, avirulent phenotype of pneumococcus to the virulent type III capsular phenotype, Hotchkiss pursued this work further and according to Horace Judson:

> After many attempts, began to find other inheritable characteristics, such as resistance to penicillin, that could be passed from pneumococcus to pneumococcus by purified transforming principle.

Harriet Ephrussi-Taylor was a formidable scientist in her own right. I mentioned in an earlier chapter how enthralled I was when I launched my own career in molecular biology in the late nineteen-sixties to discover that both she and Boris Ephrussi were on the faculty of the Biology Department at Case Western Reserve University in Cleveland. Harriet was a superb phage geneticist and a scintillating lecturer. Both Hotchkiss and Ephrussi-Taylor were quite proficient in the many subtle nuances of pneumococcal transformation and were keen on exploiting this technology as a means of investigating genetic recombination between transforming DNA and that of the recipient genome. Lacks recalled:

> Soon after joining the Hotchkiss lab I heard a lecture at Columbia University by the illustrious geneticist Guido Pontecorvo. He was mapping a gene required for adenine biosynthesis in *Aspergillus* by recombination. I was fascinated by this work and after consultation with Hotchkiss I decided to pursue studies on recombination during DNA transformation for my Ph.D. thesis.

In order to do any sort of reasonably quantitative work using bacterial transformation, it was imperative to take into account variations in the competence of the bacteria; that is, the ability of the bacteria to take up the transforming DNA and integrate it into the recipient genome, a poorly understood physiological state of the cells which varied considerably from culture to culture. Lacks related:

> We had to have an internal standard in order to correct for variations in competence, which of course was something that we couldn't control. We used the well-characterized streptomycin-resistance marker and various mutant maltose-utilization markers that I had isolated. The frequency of transformation to streptomycin resistance provided an internal standard of competence for transformation

in any given experiment. We observed that we could get recombination between some different pairs of mutant markers and not between others, and we could make sense of the recombination frequencies only if we related them to the frequency of transformation to the wild-type donor marker by itself. These markers, we found, varied widely in transformation ability. And so I began to learn something about recombination in pneumococcus.

Whilst a graduate student with Hotchkiss, Lacks participated in a special program that provided travel and short-term support for graduate students abroad. This took him to Jacques Monod's laboratory at the Pasteur Institute in Paris, where he spent a year working with François Gros. Harriet Ephrussi-Taylor was operating her own research program in Gif-sur-Yvette, just outside Paris. She was aware of Lacks's work with maltose-utilization mutants in Hotchkiss' laboratory and invited him to present a seminar to her group in Gif. "She then set out to devise a similar system," recalled Lacks. "After some initial difficulties they got another suitable transformation system going, principally through the efforts of a talented graduate student, Michel Sicard." In his later years as an independent investigator in France, Sicard went on to make notable distinguished contributions to the mismatch repair story in pneumococcus. Meanwhile, Lacks returned to the United States to complete the Ph.D. program at the Rockefeller, and shortly thereafter accepted an offer as an instructor in the Department of Biology at Harvard, where he was afforded the opportunity of establishing an independent research program. He laughingly recalled that at the time he was moving, Ephrussi-Taylor wrote to him requesting some of his mutant strains. He declined to send them then because he was very much in the middle of his own experiments and was planning further studies with these strains at Harvard. "Some years later, I had occasion to ask Harriet for some bacterial strains. She informed me that she could not give them out because her graduate student was still working with them. 'Of course you will understand,' she commented emphatically."

In the early nineteen-sixties, Lacks was recruited to a scientific staff position at the Brookhaven National Laboratory, where he has spent the remainder of his scientific career. By the mid nineteen-sixties, both Lacks and Ephrussi-Taylor and her colleagues (including Sicard) had acquired considerable information about the frequency of integration of various genetic markers following DNA transformation of different *S. pneumoniae* mutants. Both groups observed the same basic phenomenon. Lacks explained:

> When one put the same wild-type transforming DNA into different mutants, one observed that some single-site mutations were transformed at much lower frequency than others, even after correction for the competence of transformation. We knew a fair amount about the nature of the particular chemical mutagens that we used to generate mutant strains for these studies, so we had a pretty good notion of the base changes involved. It was evident that some single base changes were efficiently excluded during transformation, while others were not so well recognized and were eliminated less efficiently.

Ephrussi-Taylor and her husband Boris had by this time moved to the United States where she set up her new laboratory in the Developmental Biology Center at Case Western Reserve University. In a paper published in 1966, she wrote:

> The peculiar feature of transformation genetics [in pneumococcus] is that a given donor marker mutation transforms with an efficiency characteristic of the mutated site. In spite of this difficulty, mapping procedures have been devised and quantitative recombination studies performed.

She specifically commented on the different efficiency of integration of certain markers and, most remarkably, Ephrussi-Taylor and her colleague Thomas Gray correctly suggested that this might reflect the operation of a DNA repair process.

> It is proposed, on the basis of genetic evidence, that following essentially random exchanges between donor DNA and recipient chromosome, a *revision* [my italics] process appears to remove preferentially donor DNA sequences from the primary recombinant structure, and allow *repair* [my italics] along the chromosome template, leading to low efficiency in the genetic integration of these sites.

Borrowing from the recent demonstration of excision repair of pyrimidine dimers in *E. coli* cells exposed to UV radiation by the Setlow and Howard-Flanders laboratories, Ephrussi-Taylor and Gray elaborated on the possible nature of this DNA repair process.

> Following the discovery of excision enzymes as the agents of dark repair of ultraviolet damage in DNA, their intervention in recombination has been shown to be probable. . . . It would appear that at the stage of the formation of a primary recombinant structure, i.e., at an early stage in the reaction between donor DNA and recipient chromosome, a mechanism enters into action which destroys certain segments of either donor or recipient polynucleotide strands, showing, however, a strong preference for destroying the donor nucleotide sequence. The signal for this destruction would appear to be the configuration of the region which is heterozygous (heteroduplex) for a particular kind of mutational site. The mutations which evoke the destruction mechanism seem to be point mutations. We would like to suggest that destruction is the effect of *error-correcting enzymes* [my italics], which are recognizing single unpaired bases along a linear hybrid structure, the latter, in all likelihood, a heteroduplex composed of one strand of donor and one of recipient DNA.

Shortly after this work was published, Lacks's interests shifted towards understanding the biochemical mechanism by which transforming DNA was integrated into the genome of recipient cells. It was intuitively obvious to him that one or more deoxyribonucleases must be required to effect genetic exchanges, so he began to purify and characterize this general class of enzymes from pneumococcus. By now, the attentive reader must surely be struck by how extensively the DNA repair field profited from the widespread interest in the early and mid nineteen-sixties in understanding the biochemistry of genetic recombination. More examples of this benefit will be described later in this and in the succeeding chapter. In each case, an investigative group cogently reasoned that the key to unlocking the secrets of how exchanges between two DNA molecules were properly accomplished was in identifying the enzymes that cells must possess in order to effect these exchanges. The luxury of hindsight affords glaring recognition of the naiveté of many of the experimental approaches used to identify these enzymes, which we of course now know to be exquisitely specific when presented with the correct substrate and when tested under appropriate conditions of extract preparation. In the nineteen-sixties and even for an extended period of the nineteen-seventies, there was little if any appreciation for the extent of the repertoire of deoxyribonucleases in bacterial cells, nor for the ability of many of them to attack nonspecifically many DNA substrates. Hence, there was relatively little concern about the potential for these, as well as other enzymes that find their substrates in DNA, to confound experiments on recombination. The history of the discovery of DNA repair enzymes is replete with examples of how extensively this potential was in fact realized.

Lacks, together with his colleague Bill Greenberg, isolated and characterized two deoxyribonucleases from *Diplococcus pneumoniae*. One of these was an endonuclease and the other an exonuclease. The obvious question to Lacks was whether or not these enzymes played any role in the integration of transforming DNA. So, he set about developing a screen that would allow him efficiently to identify nuclease-defective mutant strains. In a paper published in 1970, he wrote:

> To test whether either the endonuclease or exonuclease was involved in the transformation of *D. pneumoniae* or in its ability to repair UV damage, mutants that lacked the enzymes were sought. In the process of obtaining deoxyribonuclease-deficient strains, mutants deficient in other functions possibly pertinent to the mechanism of transformation were also obtained and examined.

As luck would have it, one of these mutants was observed to have lost the phenotype of low efficiency integration of genetic markers during transformation. Lacks called these mutants (which he identified with a surprisingly high frequency) *hex*, for *h*igh *e*fficiency (*X*) integration. *X* was quaintly added to connote that the precise reason for this result was unknown. In his 1970 paper, he stated that:

> The occurrence of strains that lack the property [of low-efficiency transformation] suggests that low integration efficiency might be due to a positive, gene-determined factor, such as an enzyme. This . . . supports the sort of hypothesis advanced by Ephrussi-Taylor in which an enzyme recognizes and excises mismatched bases. The high frequency of *hex* mutants in the NG [nitrosoguanidine]-treated population suggests that loss of the *hex+* mechanism may have survival value for cells when, on account of a high mutation rate, the frequency of mismatched bases in their DNA is exceptionally high. If excision of DNA in the region of mismatched bases is involved in the *hex+* mechanism, the enzyme missing in the *hex* strains could be a deoxyribonuclease of appropriate specificity.

Like Sandy Lacks, Maurice (Maury) Fox, presently a professor of biology at the Massachusetts Institute of Technology, began his professional education in a discipline quite removed from genetics. He studied meteorology as an undergraduate at the University of Chicago and took his Ph.D. in nuclear chemistry at the same institution. Following a postdoctoral stint with the renowned physicist turned biologist Leo Szilard and his colleague Aaron Novick at the University of Chicago (whose tangential adventures with photoreactivation are recounted in Chapter 2), Fox also joined Hotchkiss's laboratory at the Rockefeller Institute, in this case as a junior faculty member. Naturally, he encountered Sandy Lacks when Sandy was a graduate student there. As a member of the Hotchkiss laboratory, albeit as an independent investigator, Fox also studied aspects of the DNA transformation process and demonstrated that transformation in pneumococcus is effected by the integration of single-stranded DNA, not duplex DNA as might have been intuitively anticipated. In the mid nineteen-sixties, Fox was invited by Walter Guild at Duke University to present a seminar on this work. Fox recounted:

> During the course of my discussions with Guild it became apparent that a genetic marker which in my hands integrated with low efficiency, did exactly the opposite in a strain that Guild had. So I got hold of Guild's strain and when I returned to the Rockefeller I compared the two strains directly and indeed they gave opposite results. That experiment convinced me that a genetic determinant specified integration efficiency during pneumococcal transformation. In fact Guild's strain turned out to be carrying a mutation in the gene that came to be called *hex* by Lacks.

When recalling his studies that led to the formal demonstration of genetic determinants for mismatch repair in *S. pneumoniae*, Lacks credits "a little assist from Maury." He recalled that, "We talked about these results and he definitely influenced me to think that the phenotype of the loss of low efficiency integration might be the result of a cellular mutation." In the paper in which he described the identification of his *hex* mutants, Lacks wrote the following:

> Guild and Fox (personal communication) found that the same *nov* marker that transforms [strain] R6 with low efficiency will transform Rx, the strain used in Guild's laboratory, with high efficiency. Markers in the amylomaltase locus that gave low-efficiency integration in the transformation of mutants derived from R6 gave high-efficiency integration after the corresponding mutations were introduced into Rx. This confirmed a suggestion by Fox (personal communication) that the integration of all markers with high efficiency was characteristic of [Guild's] recipient strain. Similar strains of genotype designated *hex*, were isolated by individually testing clones of an NG [nitrosoguanidine]-treated culture for transformation with normally high- and low-efficiency markers. Hex mutants appear to have lost the ability to give low-efficiency transformation, an ability that is a property of the wild-type strain. The *hex+* character was transformed back into one such mutant with a frequency sufficiently high to indicate that mutation of a single gene caused loss of the property in question.

Maury Fox subsequently enjoyed the collaboration of a talented postdoctoral fellow, Gerard Tiraby, who received his formal graduate training in France under the expert tutelage of Michel Sicard (the former graduate student with Ephrussi-Taylor). Tiraby and Fox found that the genetic determinant that influenced integration efficiency in pneumococcus also influenced the spontaneous mutation frequency. Hence, by the early nineteen-seventies, Fox, Tiraby, and Sicard recognized that the *hex* mutant was a spontaneous mutator, the precise phenotype expected of cells defective in the repair of mismatched bases. In a paper published in 1973, Tiraby and Fox stated that:

> The loss of the hex+ function results both in a loss of the capacity to eliminate the low-efficiency markers in transformation and a substantial increase in the spontaneous mutation rate. These properties of the hex- strain could result from the loss of a capacity to eliminate certain classes of mismatched base pairs that occur as intermediates in both transformation and mutagenesis.

The reader will recall from earlier chapters that the famous experiment by Avery, MacLeod, and McCarty, which is formally acknowledged as the first definitive demonstration that DNA is the genetic material of cells, utilized transformation of pneumococcus. Miroslav Radman informed me that retrospective analysis established that the genetic marker scored in their transformation experiments carried a deletion that was stably integrated with high efficiency. Of course, Avery and his collaborators had no way of knowing this, nor did they have reason to care. Had these pioneers of molecular biology unknowingly utilized a marker with a single base mismatch, the efficiency of its integration might have been considerably reduced and transformation of the type II serological marker might have been missed. Had such a situation transpired, one cannot help but wonder how and when the formal discovery that DNA was the genetic material would have occurred, and how different the history of molecular biology might have been.

The experimental evidence provided by Setlow and Carrier and Boyce and Howard-Flanders that damaged bases can be excised from the genome also fired the imagination of other scientists who were seeking biochemical explanations for aberrations in recombination. This group included Robin Holliday, then director of the Medical Research Council (MRC) Unit at Mill Hill, London, where his laboratories were situated next door to those of John Cairns. I have been acquainted with Robin Holliday for a number of years, though never as intimately as I would have liked had time and circumstances permitted. I found him to be an extremely affable man whose warm smile and charming sense of humor always put me at ease. But this low-key, "laid-back" demeanor belied a formidable intellect and a finely honed critical aptitude which he had little hesitancy in unleashing as he deemed appropriate. Robin Holliday did not tolerate fools gladly! His scientific interests embraced many areas of molecular biology and genetics, in particular recombination, DNA repair, and the biology of aging. Like many of his contemporaries, he was a potent theoretician and his ideas and hypotheses stimulated many experiments in laboratories around the world. His conceptual contributions to genetics included a model of a four-stranded recombination intermediate which now bears his name as the "Holliday junction."

In 1964, the year that the seminal excision repair manuscripts were published, Holliday wrote a paper in which he speculated about the mechanism of gene conversion. Gene conversion is a term that describes the nonreciprocal transfer of genetic information from one DNA molecule to another during meiosis, thereby resulting in deviations from strict Mendelian segregation of genetic markers. The phenomenon was well described in yeasts and other fungi in which the segregation of genetic markers could be accurately (and rather simply) followed by examining the genotype of spores generated during meiosis. So, for example, in an organism that normally elaborates eight spores as products of meiosis, one might very occasionally observe a 6:2 wild-type:mutant ratio of segregation of a genetic marker of interest, rather than the expected 4:4 ratio, suggesting that some of the mutant alleles had been "converted" to the opposite type. In fact, it was precisely this type of observation in the fungus *Bombardia lunata* by H. Zickler in 1934 that led to the first clear demonstration of gene conversion.

Further study of the phenomenon led to the recognition that the process of gene conversion was associated with crossing-over events during meiosis, that is, with recombination. Just a year prior to Holliday's paper, H. L. K. Whitehouse proposed that a mutant allele of a gene on one chromosome might be generated by heterozygosity of DNA during the process of recombination, and suggested that instead of this allele undergoing normal meiotic segregation, the region of heterozygosity might be altered such that the mutant allele was converted to the wild-type form.

A fundamental feature of the theory of crossing-over that has been proposed here is the occurrence of hybrid DNA in the region of cross-overs or potential cross-overs. Heterozygous DNA would imply segregation of genetic differences at the first mitosis after meiosis, and such an occurrence, although known, is quite rare. It is presumed, therefore, that the hybrid DNA, although often potentially heterozygous, usually becomes homozygous in the process of formation of the hybrid molecules. This could be achieved by the removal from one chain at each heterozygous site or region of those nucleotides which did not conform to those in the other chain, followed by the insertion of nucleotides complementary to those in the first chain.

Robin Holliday took this model a step further in suggesting that "if there are enzymes which can repair points of damage in DNA, it would seem possible that the same enzymes could recognize the abnormality of base pairing, and by exchange reactions rectify this."

Holliday and Harriet Ephrussi-Taylor were correct in their assertions that mispaired bases are rectified by an excision repair mode. However, this excision repair mode is genetically and biochemically quite distinct from the nucleotide excision repair system that deals with pyrimidine dimers and various types of chemical damage in DNA, and from the base excision repair mode initiated by DNA glycosylases. Nature evolved a unique and specific set of proteins to deal with this type of genetic instability. For the convenience of the uninitiated reader, it needs to be stated here that a fundamental problem that must be solved in all organisms engaged in mismatch repair is the discrimination between the correct base and the incorrect base in a mispaired set. It is difficult to conceive of how nature could have solved this problem by a mechanism that relied on the exclusive recognition of the mispaired bases. For how then could cells differentiate between a C·A mispair with C as the correct base and A as the incorrect one, and the same mispair in which the opposite was true? The solution that nature evolved was a two-pronged recognition mechanism. One protein (or set of proteins) is dedicated to the recognition of mispaired bases per se, regardless of the appropriate or inappropriate residence of the bases in either of the two DNA strands. A second protein (or protein complex) is dedicated to the specific identification of the DNA strand carrying the incorrect base. Since mispaired bases most often arise during the synthesis of new DNA strands, this strand-specific identification process borrowed from biochemical events that distinguish daughter DNA strands from their mothers during DNA replication. In *E. coli*, in which the mechanism of strand discrimination during mismatch repair was first elucidated, and in related prokaryotes, the A residues in all 5'GATC3' tetranucleotide sequences are methylated by a specific methylase. The biological significance of this methylation is not relevant to the present discussion and probably did not evolve for its specific utility in mismatch repair. This methylation occurs very soon after a 5'GATC3' sequence is replicated. But there is a brief kinetic window during which the newly-synthesized DNA strand near the replication fork is not methylated, thereby defining a newly-replicated segment of double-stranded DNA that is transiently hemimethylated. Hemimethylated DNA is the substrate for the second prong of the two-pronged recognition system. Specific proteins bind to the nonmethylated strand at hemimethylated sites, thereby flagging that strand for excision repair at the mismatch sites. It took many years to reveal the proteins in *E. coli* that specifically recognize hemimethylated DNA and the genes that encode them. Indeed, in a comprehensive review of DNA repair enzymes written as late as 1982, Tomas Lindahl correctly stated that "little is presently known about the biochemistry of mismatch repair, and this process may be regarded as a major remaining mystery of DNA metabolism."

The history of the discovery of the repair of mispaired nucleotides in *E. coli*, in particular the solution to the strand discrimination problem, is primarily the story of the evolution of Matthew Meselson's eclectic scientific interests. Like his contemporary and former collaborator Frank Stahl, Matt Meselson was one of the primary makers of the revolution in biology that rocked science in the nineteen-fifties and early nineteen-sixties. At the tender age of 16, Meselson had accumulated enough credits to graduate from the high school he attended in southern California, where he was born and raised. Upon confronting the school registrar with this information, he had his first memorable introduction to the harsh realities of bureaucracy. He was informed that his stellar academic performance notwithstanding, he simply could not graduate because he required a further year of credit in physical education! "That astounded me," he recounted, "because up until then I had thought that my elders, if not infinitely wise, at least knew what they were doing in guiding my education. I'd simply follow the track they laid out and everything would be fine." Emphatically disinclined to devote a year of his life to athletic activities, Meselson learned about and elected to apply to a program at the University of Chicago for advanced high school students; the same program that admitted James Watson through its doors at the age of 15 in 1943. Three years at the University of Chicago broadened his intellectual horizons, and as new vistas of opportunity unfolded for the romantic young Meselson, his firm childhood conviction that he would become a chemist was seriously challenged. Meselson confided with proverbial tongue in cheek:

> The University of Chicago experience led me to the decision that I should become a psychoanalyst and apply psychoanalytical theory to the behavior of nations and in that way try to contribute to peace in the world. I was seriously confused, so in 1949 I decided to go to Europe for a year, which I spent mainly in France. I tried to get into Russia because I was curious about these people who were once our allies and now were apparently our enemies, but I was denied access to the country. While in Europe, I came to the understanding that while philosophy, psychoanalytic theory, and sociology were certainly appropriate vehicles to explore important problems in the world, they were too vague; intellectually too soft to dedicate a professional career to. I realized that I didn't want to devote my life to a discipline which might force me to the conclusion at the end of my career that much of what I had contributed was perhaps factually inaccurate. I finally recognized and accepted that I really wanted to pursue a life's work that would yield some basic truths, even though these might have limited significance in the greater scheme of things.

So, Meselson decided to join the world of objectivists and become a scientist. But those who have followed his career are aware that Meselson never fully abandoned his romantic aspirations. In the later stages of his career in particular, he effectively translated his considerable scientific prominence in the energetic pursuit of peace in the world through a variety of formal and informal mechanisms.

Following his return from Europe, Meselson returned to Chicago to complete a further year of study in chemistry and departed armed with a letter from the University stating that although it did not formally offer a bachelor degree in chemistry, if it did so it would certainly confer such a degree to one Matthew Meselson! He journeyed back to his native California and in 1953, the year that Watson and Crick published the structure of DNA, he enrolled for graduate studies in biophysics at the University of California at Berkeley. Meselson recalled:

> I expected that I would encounter a lot of interesting biophysics there. But much to my disappointment I did not. I became a graduate student with John Gothlin who soon recognized that I was seriously misplaced at Berkeley and suggested that the person best suited to mentor my interests was Linus Pauling at Caltech.

Meselson was familiar with the Pauling family through his friendship with Linus Pauling's son, Peter. However, he was not entirely convinced that he was adequately qualified to study with the "great Pauling." "So," he recounted, "I decided to return to the University of Chicago and become a graduate student with Nicholas Rutchevsky, a Russian scientist who applied second-order differential equations to various boundary conditions in pseudo-biological situations. I was accepted at Chicago and everything was set."

The summer before he was due to depart for Chicago, Meselson was invited by his friend Peter Pauling to a swimming party at the Pauling residence. This social occasion turned out to be a major crossroad in his life. Meselson recounted:

> I was in the water splashing around when Linus came out, all dressed up as I recall, including a vest and necktie. He knew me a little and had followed my undergraduate career. He stared down at me in the pool and asked what I was going to be doing after the summer. When I told him where I was going and who I proposed to work with he appeared absolutely dumbfounded; a look that I've never quite seen since. He removed his glasses and said, "Matt, that's a lot of baloney. Why don't you come to Caltech and be my graduate student?" So half naked and dripping with water I looked up at him, fully dressed, the world's greatest chemist, and I simply said, "OK sir, I will."

In Pauling's laboratory, Meselson enthusiastically immersed himself in X-ray crystallography.

> I didn't like pure biology at all at that time of my life. The biology I had been exposed to at school was the "creepy" and "wet" kind and you had to memorize everything! I far preferred chemistry and physics and mathematics. But I really very much wanted to apply these quantitative disciplines to some sort of biological problem.

Pauling initially proposed that Meselson solve the crystal structure of tellurium.

> He teased me with the admonition that chemists who worked with tellurium often got what he called "tellurium breath," a foul odor which led to their immediate ostracism from society. He told me that some tellurium fanatics had even committed suicide!

Meselson insisted to Pauling that he wanted to understand the structure of "something that was comprised of carbons, hydrogens, and nitrogens, something organic, so that I could learn some real biology." So, he was challenged with solving the structure of some esoteric dipeptide, the name of which is not at all relevant to this discussion, and this work constituted much of his Ph.D. thesis.

Sometime in early 1954, while still a graduate student of course, Meselson formulated an idea that would eventually lead to his famous density gradient experiments with Frank Stahl, which provided the first definitive demonstration of the semiconservative nature of DNA replication. He attended a seminar at Caltech from the French molecular biologist Jacques Monod about his (then early) studies on the induction of expression of β-galactosidase in *E. coli*; experiments which, when finally consummated with his primary collaborator François Jacob, would have profound significance for the essential features of gene regulation in bacteria as we understand this process today. The central question at the time was still the nature of the induction process. Meselson recalls that Monod proposed carrying out "some sort of horrible experiment based on changing the osmotic pressure in the cell in order to determine whether the enzyme β-galactosidase was activated from a dormant form or was really synthesized de novo during induction." At that time, Meselson was deeply involved in Pauling's famous Caltech course on the nature of the chemical bond, and was intrigued by what he had learned about the use of heavy isotopes for measuring chemical bond energies.

> I had all this heavy isotope stuff floating around in my head, and while listening to Monod the idea suddenly occurred to me that the right way to test for the induction of β-galactosidase would be to grow bacteria in heavy water, add galactose to induce the enzyme, and then put the bacteria back into light water, or vice versa. One could then open the cells and centrifuge the lysate in a solution that had a density exactly equal to that of half heavy/half light β-galactosidase. If the enzyme was all light, it would stay at the top of the gradient and if it were all heavy, it would go to the bottom. I had no notion of a density *gradient* at that time. My thought was that the enzyme would either go up or down and where it appeared would tell you whether it was made of light or heavy precursors.

The more he thought about this experiment the more he liked it and so he excitedly related his idea to Pauling, who patronizingly told him to "finish your crystal structure and get your Ph.D.!"

This gentle brush-off prompted Meselson to seek out the fearsome Max Delbrück! He laughingly recalled:

> I was nervous of course, because Delbrück's reputation among the graduate students at Caltech was legendary. Additionally, I had roomed with a fellow graduate student named Martin Karplis and he and Delbrück had recently had a major falling-out. But eventually I made an appointment to see him. I was supposed to know something about structural chemistry at this stage of my career, so when I entered his office the first thing that Delbrück said to me was "What do you think of this new structure of DNA?" To which my dumbfounded reply was something to the effect that I knew nothing about it! Max rose from his chair and reached behind him to a shelf where he had a stack of reprints and preprints which he had obtained from Jim Watson. He threw them at me and chastisingly asked me "How could it be that you don't know the most important development in biology in the last 10 years? Read this and don't come back until you've understood it all." Well, I was suitably chastised, but I was also excited because what really remained with

me from that abortive conversation was the "come back" part. After I read these papers, I excitedly came to the realization that it was not β-galactosidase induction that wanted my density-shift experiment, it was DNA replication!

A short time later, Meselson met Jim Watson at Caltech and related his idea for investigating the mechanism of DNA replication. Watson liked the notion immediately but also advised Meselson to "wait until your thesis in crystallography is done and then go to Sweden to do the experiment." According to Meselson, Watson's recollection is that he suggested Sweden because there were plenty of women there! Meselson's recollection is that Watson suggested Sweden because there were plenty of centrifuges in Sweden. "Maybe it was both," Meselson laughed. "If so, it's very telling as to which of us remembers what, and I'm ashamed for thinking it was the centrifuges!"

Watson invited Meselson to spend the summer of 1954 at the Marine Biological Laboratory in Woods Hole, Massachusetts to assist with the teaching of a summer course in general physiology that Watson, Crick, and Brenner were involved with and to help out with experiments that Watson was engaged in. Like Cold Spring Harbor on Long Island, Woods Hole had developed a reputation as an excellent place to spend the summer doing research and taking advanced courses. And this is where Meselson and Stahl (who was taking the physiology course) first met. Meselson related:.

> One day during that summer, Jim [Watson], Francis [Crick], Sydney [Brenner], and I were standing around in one of the rooms in the brick building at Woods Hole, when Jim pointed out of the window across the street. "See that guy down there?" he asked. "That's Frank Stahl. He's a real smart ass! He thinks he can do anything and we're going to test him to see whether he can do the entire Hershey-Chase blender experiment in a single lab session!" I decided that Stahl sounded like an interesting person to meet so I wandered down and there was Frank sitting up against a tree on the lawn selling gin and tonics to any interested takers and simultaneously trying to integrate a mathematical function that he'd put together. He explained to me that this was a problem in phage radiation genetics which involved a complicated series of integrals. At that time, I knew more mathematics than Frank, so I quickly wrote down the correct equations—and that was the beginning of our friendship. During our discussions that summer, I told him about the DNA experiment that I had in mind. As it turned out, Frank had already arranged to come to Caltech for a postdoctoral fellowship the following year, so we agreed that I would be a "good boy" and finish up my X-ray crystallography as everyone (including Stahl) insistently advised, and that when he came to Caltech we'd try this experiment together. When he finally did come out there, he and Jan Drake and I lived together for a while. I was cook, Frank was the dishwasher, and Jan was the gardener. Frank remained adamant that I should finish up my crystallography work before we did the experiment. He said it would be bad for my character if I didn't, and he really meant it! So at a very definite point in time I finished my thesis and bang, we did the density gradient experiment.

A comprehensive account of this experiment can be found in Horace Judson's book *The Eighth Day of Creation*.

This first serious foray into the world of biology converted Meselson from a chemist to a molecular geneticist, despite the fact that he never took a single genetics course from any of the numerous genetics professors who taught at Caltech when he was a graduate student, a fact that he now somewhat laments. As he deepened his interest in this emerging discipline, and in DNA replication in particular, he encountered in the literature a hypothesis that recombination might involve a specialized form of DNA replication called "copy choice replication," during which the putative DNA replication machinery literally jumped from a template strand in one chromosome to that in the homologous chromosome, thereby essentially recombining genetic information from two chromosomes. The idea is historically credited to John Belling, who suggested this mechanism with respect to the duplication of chromosomes in the late nineteen-twenties. This notion did not sit well with Meselson's firmly ingrained concept of the strictly semiconservative nature of DNA replication, and both he and Stahl became increasingly interested in understanding the true mechanism of recombination, the alternative view of which was that it took place by the breakage and rejoining of DNA molecules. Indeed, Stahl went on to devote the remainder of his own highly distinguished scientific career to deciphering this aspect of DNA metabolism. Meselson's considerable expertise with density gradient centrifugation quickly led him to the realization that the technique might be profitably exploited to examine whether or not recombination in phages involved breakage and rejoining of DNA. So, when Stahl left Caltech, Meselson collaborated with Jean Weigle, a well-known member of the (rapidly expanding) phage λ club. Some of Weigle's studies with this phage will be explored in a different context in the next chapter on the discovery of the SOS phenomenon. This collaboration between Meselson and Weigle yielded a series of classic experiments that formally documented recombination of phage λ DNA by the process of breakage and rejoining.

This text is not the appropriate vehicle to document the details of Meselson's extensive theoretical and experimental contributions to our understanding of genetic recombination. For the purposes of the present discussion, it suffices to note that at some point in his grapplings with the molecular mechanism of this process in phage λ he came to the realization that he required a biological (as opposed to a physical) yardstick that could distinguish between two different recombinant λ DNA molecules. His solution to this problem was to attempt to exploit the rapidly developing elucidation of the phenomenon of host-controlled modification and restriction of DNA. This phenomenon functions as a type of immune surveillance system in bacteria which enables a bacterial cell to discriminate between its own DNA and that of an intruder, such as that from an invading bacteriophage. Discrimination between "self" and "nonself" is achieved by enzymatic modification of "self" DNA such that it is rendered immune to degradation by restriction endonucleases that specifically recognize and cut unmodified DNA. Modification usually takes the form of methylation of the 6-amino group of adenine residues in a specific DNA sequence, typically 4–6 nucleotides in length. As every student of biology now knows, multiple such sequences have evolved in nature for this purpose, each of which, when unmethylated, provides a substrate for a different restriction endonuclease. In some organisms, the modification and restriction activities are built into the same enzyme; so-called type I restriction enzymes. Other organisms are endowed with distinct methylating and restriction endonuclease enzymes; so-called type II restriction enzymes. It is, of course, the latter class of restriction enzymes and the recognition, principally by Dan Nathans and his colleagues in the early nineteen-seventies, of their extreme utility for cutting DNA faithfully at unique sites that brought about a second revolution in biology; the era of recombinant DNA technology. This revolution, which in keeping with Horace Judson's creationary theme might be dubbed "the ninth day of creation," followed immediately on the heels of the revolution ushered by Watson and Crick's observations on the structure of DNA just 20 years earlier. This newest revolution in biology shows little signs of decline as we enter the 21st century.

Host modification of phage λ was discovered by his collaborator and colleague at Caltech, Jean Weigle, so Meselson was very close to the source of this discovery and intimately familiar with the biology of the phenomenon as it evolved in various laboratories. At that time, it was not known that modification occurred by DNA methylation, but as we shall presently see, when this fact was ultimately established, it profoundly influenced Meselson's later thinking about the role of DNA methylation in mismatch repair in E. coli. Meanwhile, having crossed the country to Harvard University, where he joined the faculty and spent the rest of his career as a scientist (and social activist), Meselson completed the experiments that ultimately convinced him that recombination in phages and in fungi (the other traditional stronghold of the copy choice model) did indeed transpire by breakage and rejoining. Meselson related:

> I was like an errant knight on my horse wanting to throw my lance through copy choice and destroy it completely because its existence anywhere in biology was a threat to our little world. So I was forced to explain the aberrant segregations [gene conversions] that one occasionally observed during recombination by breakage and rejoining. At Caltech, Jean [Weigle] and I had carried out density gradient experiments to ask whether there was any DNA replication at all associated with phage recombination. We observed phages that were almost completely "heavy" but their density was shifted just a little bit towards the light end. *This gave me the idea that some sort of enzymatic excision repair in a region of hybrid DNA overlap might explain gene conversion* [my italics].

In 1963, Meselson presented a paper at the 16th International Congress of Zoology. In the proceedings of the meeting, which were not published until two years later under the title *Ideas in Modern Biology*, he wrote:

> Although conclusive evidence has not yet been obtained, certain physical observations which we will not describe suggest rather strongly that the overlap in recombinant bacteriophage chromosomes contains only two DNA strands. . . . Still another characteristic of the joining of DNA molecules suggested in the course of the above-mentioned observations is the removal and resynthesis of a small amount of DNA in the course or recombinant formation. . . . The possibility that some DNA within the overlap is removed and resynthesized suggests an explanation for certain aberrant segregations observable when all the products of meiosis are recovered as is possible, for example, in various fungi. . . . Such aberrant segregations are indeed known to occur and to be correlated with recombination of markers nearby. They have been termed gene conversion to indicate the possibility that they result from a process other than normal recombination. However, there is no compelling evidence against the simpler view that gene conversions are an expression of the removal and resynthesis of DNA associated with normal recombination as suggested here. Since this symposium was given a similar explanation has been presented by Whitehouse.

It is really quite remarkable how influential the basic observations of excision repair of UV-irradiated DNA by Dick Setlow and Paul Howard-Flanders turned out to be. As documented several times in this chapter, the discovery that bacterial cells possessed the basic wherewithal to cut damaged segments out of DNA inspired a number of investigators, including Harriet Ephrussi-Taylor, Sandy Lacks, Robin Holliday, and Matt Meselson to borrow (correctly) from this paradigm in extrapolating to the repair of mismatched nucleotides. His longstanding abhorrence of biochemistry (especially enzymology) notwithstanding, Meselson recognized that it was imperative to demonstrate the existence in cells of enzymes required for the breakage and rejoining of DNA during genetic recombination. So, like several investigators at the time, he set up experiments to determine whether one could detect the breakage and rejoining of phage DNA molecules that were differentially radiolabeled, after incubating them with extracts of *E. coli*. He recalled:

> I was a complete flop at that sort of thing. Nothing was coming out of these experiments. Eventually, in desperation, I decided that we had to get something useful out of them. "To hell with recombination," I thought. Since we had these various labeled phages lying around, I decided to use them to investigate the enzymology of host-controlled restriction and modification instead.

So, Meselson set up a sucrose gradient sedimentation experiment to begin his search for enzymes involved in host-controlled restriction. For this experiment, he used λ DNA from phages grown in a strain of *E. coli* nonrestrictive for phage P1 and labeled with one radioisotope, and λ DNA from phages grown in a different strain restrictive for P1 DNA and labeled with a different radioisotope. These DNA preparations were incubated with an extract of either the restrictive or the nonrestrictive strain with the hope that the restrictive extract would selectively degrade the DNA of only the nonmodified phage. A third incubation was simply a control with no added extract. Meselson lamented:

> But the centrifuge rotor had slots for six tubes. So I decided to occupy the other three slots by repeating the identical three reactions with some sort of possible cofactor for the putative restriction reaction and quite naively elected to add ATP. Only one tube showed any change in the distribution of the radiolabels. It was the reaction with the restricting extract plus ATP. We immediately realized that we had probably discovered an activity that was responsible for host-controlled restriction and we decided to go hell-bent for it.

Meselson and his colleagues went on to purify a restriction endonuclease from *E. coli* that was also endowed with modification activity; a type I restriction endonuclease. On the basis of its demonstrated requirement for a methyl donor for the modification reaction, they speculated in a paper that was not published until 1968 that host modification involved the methylation of DNA. Tom Lindahl recalls hearing Meselson present his new unpublished and preliminary data on the ATP-dependent restriction endonuclease in *E. coli* strain K12 at a Gordon Conference in the late nineteen-sixties. "It was perhaps the most exciting talk at the meeting," Lindahl remarked to me some 28 years later.

Following this brief but, as you will presently see, historically significant diversion into DNA methylation, Meselson returned to the question of how gene conversions might occur during recombination and their relationship to a possible DNA excision repair mode. In the early nineteen-seventies, Judy Wildenberg, an experienced fungal geneticist, joined Meselson's laboratory at Harvard and began experiments with λ DNA heteroduplexes containing deliberately engineered gene conversions, that is, mispaired bases. When these molecules were introduced into *E. coli*, genetic analysis of the progeny phage unambiguously revealed that the mismatched regions had undergone repair. These results were published in 1975. A year later, Meselson published the results of further studies on mismatch repair, this time with his new graduate student, Robert E. Wagner, Jr. It was in this paper that Meselson's familiarity with the phenomenon of DNA methylation lent itself to a prophetic insight about the correction of mismatches generated during DNA replication. He and Wagner wrote that:

> The biological role of mismatch repair remains a matter for speculation. Depending on the parameters of heteroduplex formation during genetic recombination and on the pattern of repair excision, mismatch repair may make a major contribution to the overall recombination frequency of very close markers. However, in view of the uncertainty regarding the biological role of recombination other possible functions of mismatch repair should be considered. *For example, mismatch repair may act to correct mutations that arise as replication errors. If so, it may be that mismatch repair acts* in a [strand] directed manner in conjunction with sister chromatid exchange, or that it occurs *with particularly high efficiency on newly synthesized DNA strands, possibly because of their undermethylation* or because of a special relation to the replication apparatus [my italics].

We will shortly encounter Miroslav Radman's disappointing initial involvement in the mismatch repair story when he was a postdoctoral fellow in Meselson's laboratory in the early nineteen-seventies. In the summer of 1975, Radman, Meselson, and others were attending an annual international meeting on genetic recombination in Nethybridge, Scotland. These meetings, dedicated specifically to the topic of recombination, started out in Aviemore, Scotland before moving to Nethybridge and more recently to France. Radman recalls that at the 1975 meeting "there was lots of discussion and speculation about mismatch repair and its role in correcting replicational errors." In particular, it was now well recognized that the *hex* mutants of *Streptococcus pneumoniae* were spontaneous mutators and hence were likely to be defective in postreplicative mismatch correction. Radman recalled:

> There was discussion about the possibility that mismatch correction might occur immediately after DNA replication. But if this was the case, what directed the repair to the newly synthesized strand? Maury Fox and Robin Holliday and possibly others who were there at the time were convinced that it had something to do with the fact that newly synthesized DNA strands had discontinuities.

For *Streptococcus pneumoniae* this indeed turned out to be the case. But as already indicated, *E. coli* uses a special device for providing strand discrimination during mismatch repair. Radman reflected:

> I don't recall whether or not Matt said this publicly at the meeting, but I vividly remember that some of us were sitting around one night sipping the fine Scotch malts they had at the hotel in Nethybridge and Matt turned to me and said, "Miro, if I were a mismatch repair enzyme, I would look for sites in DNA that were not methylated," or something to that effect. Literally within an instant of Matt saying

this to me I flashed on a paper that I had read about a month earlier by Martin Marinus and his colleague Ronald Morris on the isolation of a mutant strain of *E. coli* that they called *dam*, for DNA *m*ethylation. Marinus and Morris had shown that this mutant was defective in its ability to methylate DNA and was a mutator strain. So, I said to him, "Matt, as I recall it has recently been shown that a DNA methylation mutant is a spontaneous mutator. You have to look at this paper, it was just published." Matt rather nonchalantly replied, "That's interesting. That's exactly what one would predict if there was a loss of strand discrimination."

Sandy Lacks pointed out to me that although the strategy of DNA methylation for strand discrimination may indeed have had its conceptual birth at the 1975 Nethybridge meeting, the notion of strand breaks as a key factor for strand discrimination during mismatch repair evolved earlier.

> My recollection is that discussions about strand breaks occurred at the 1973, 1974, and 1975 Wind River transformation meetings and at the 1974 Gatlinburg symposium on recombination. Discussions of how cells solve the problem of strand discrimination that involved Maury [Fox], Walter [Guild], Robin Holliday, Matt Meselson, Carl Tiraby, Hotchkiss and myself, among others, at the 1974 Gatlinburg meeting may have been continued by some of these individuals with Miro [Radman] at Nethybridge in 1975.

"By 1976, I was convinced that the basis for strand discrimination was DNA methylation," Meselson recounted. "I remember telling this to Jim Watson and he replied 'nonsense, it's proteins that do it.'" Meselson acknowledged that this hunch clearly derived from his intimate familiarity with methylation as a mechanism used in nature for modifying DNA, and also laughingly acknowledged that this familiarity in turn was in a sense the outcome of almost accidentally adding ATP to his modification-restriction reactions, prompted purely by his need to fill all six tube holders in an ultracentrifuge rotor.

> As I look back on those early experiments, I certainly didn't know enough biochemistry to guess intelligently about what to add as a possible cofactor, except ATP. It was like the old joke of adding "pH." In those days, phage geneticists perversely prided themselves on not knowing much biochemistry and at Cold Spring Harbor Symposia we would sometimes jokingly ask a speaker who was trying to solve some sort of biochemical problem, "Did you try adding pH?"

In a review written in the late nineteen-eighties, Meselson revisited this fundamental paradigm switch of the association of mismatch repair with recombination to its association with DNA replication.

> Although mismatch repair clearly is involved in a number of phenomena of genetic recombination, this does not tell us why such repair exists, that is, what factors selected for its emergence and maintenance in evolution. It might be asked which, if any, of its possible effects on genetic recombination are of adaptive value to the organism. It may be thought, for example, that enhanced recombination of close markers, resulting from certain patterns of mismatch repair, confers a selective advantage. A problem with this explanation is that it is not clear that this enhancement of recombination would be advantageous. . . . Uncertainty about the adaptive value of its consequences led us to consider a different possible role for mismatch repair, one which does not derive from its effects on genetic recombination.
> Mismatch repair could be advantageous if it occurred mainly on newly synthesized DNA chains, thereby correcting errors of replication. This requires some way of

recognizing which of the two chains is the copy. The DNA of most organisms is methylated. Since this is a secondary modification, the new chain will be transiently undermethylated, providing a possible basis for its recognition.

As I just mentioned, Miroslav Radman, whose contributions to other pathways by which prokaryotic cells respond to DNA damage are recounted in the next chapter, was intimately involved with several aspects of the emergence of the mismatch repair story in Meselson's laboratory, where he trained as a postdoctoral fellow in the early nineteen-seventies. Following a brief postdoctoral stint with Raymond Devoret at the Curie Institute in Gif-sur-Yvette, just outside Paris, Radman left for the scientific nirvana of the United States that he had heard so much about. Radman said:

> I had decided that I wanted to work in recombination. I loved the field of mutagenesis and DNA repair, but I felt intellectually slighted in some way by being frequently identified with the school of classical radiobiology. All these brilliant guys that I'd heard about were working on transcription and replication and recombination and DNA topology and they were members of another club to which I felt I didn't have access. I had heard a lot about Matt Meselson from Ann Rahler, his former graduate student, and I was very impressed by reading his papers and by his sense of scientific perfectionism. He was doing exciting experiments on recombination using a combination of physical and genetic approaches and I wanted very much to work with him.

As we will see in the next chapter, when Radman arrived in Meselson's laboratory at Harvard in 1970 his head was filled with cartoons and budding hypotheses concerning the mechanism of UV-radiation-induced mutagenesis in *E. coli*. But he had to put all this aside since the intellectual currency being traded in Meselson's laboratory was ideas about recombination, not mutagenesis. When Radman arrived, Judy Wildenberg's experiments with λ heteroduplex molecules were under a full head of steam. "Out of these studies a single paper emerged by Judy and Matt, which is a classic," Radman told me. "It is the single most dense paper I have ever read with respect to data and data interpretation." Radman considers Judy Wildenberg to be one of the sharpest minds in genetics that he has encountered, a view that is shared by many prominent figures in the field. Regrettably for the world of genetics, Judy Wildenberg was afflicted with multiple sclerosis and had to retire from active research in the early nineteen-seventies.

By this time, Meselson had begun to evolve the notion discussed earlier that gene conversion was associated with some sort of excision repair mode that corrected mismatches in heteroduplex DNA and Radman was urged to develop his biochemical skills and go in search of mismatch repair enzymes. The initial idea was to use velocity sedimentation in sucrose gradients to look for endonucleases that preferentially cut radiolabeled heteroduplex DNA molecules, very much along the technical lines that Meselson had used earlier to explore the enzymology of recombination and more recently (and more successfully) that of host-controlled restriction. Radman recalled painfully:

> Every experiment consumed three alkaline sucrose runs in the centrifuge. Eighteen tubes, six per rotor, were the most I could manage in a day, running the centrifuge literally night and day. Those three years of running gradients in Matt's lab were the most hands-on lab work that I have ever done in my life. It exceeds the remainder of my total scientific career since then.

And with his typical Slavic humor he laughingly rejoined that he loved the general ambiance of Harvard and the experience of working and living in that environment so much that "never since has the quality of my life been so good in the sustained absence of any positive experimental results! As Matt once told me, 'This is a character builder, Miro.'"

What kept Radman tied to the centrifuge and the cold room for these three long years was the fact that he did, with significant frequency, observe a hint of an activity that preferentially cut heteroduplex DNA; just the slightest shift in the sedimentation profile of the correct substrate DNA. But this hint never blossomed into a result that was convincing enough to persuade him (or anyone else) that it could be used as an assay to purify a putative mismatch repair endonuclease. In retrospect, he almost certainly was observing mismatch repair in operation. But what was consistently and depressingly interpreted as the instability of the putative enzyme that was teasing him, or its low specific activity, turned out in retrospect to be a problem of substrate limitation. At this time, no one yet knew that the critical secret of the *E. coli* mismatch excision repair system lay in the pattern of postreplicative methylation at 5'GATC3' sequences in each of the two DNA strands. And the DNA that Radman was using as a substrate was appropriately hemimethylated only to a very limited extent.

During Radman's final year in Boston, a graduate student joined Meselson's laboratory and thus began Radman's long and close collaboration and friendship with Robert E. Wagner, Jr., a talented young investigator who very soon after completing his Ph.D. decided that life is too short to be devoted entirely to science, especially when there are also sheep to be farmed, a passion that he indulges to this day. So, Wagner now manages a sheep farm for half of his time and provides the scientific community with the benefits of his considerable research talents for the other half. When Wagner joined Meselson's laboratory, Judy Wildenberg was still carrying out genetic analyses of gene conversions using randomly prepared heteroduplex DNA. Meselson related:

> We weren't yet able to make pure populations of Watson/heavy and Crick/light strands. We had Watson/heavy:Crick/light heteroduplexes mixed with Watson/light:Crick/heavy ones. One could still get clear proof of the occurrence of excision tracts by co-correction of spaced genetic markers, but we really wanted to nail this down biochemically.

But, by the time that Wagner joined the laboratory, the technology for cleanly separating and storing Watson and Crick strands had matured, and Wagner was able to make pure populations of Watson/heavy:Crick/light heteroduplexes or vice versa. Meanwhile, Radman had become more and more intrigued with the notion of methylation as the basis for strand discrimination during mismatch correction, the idea that had surfaced at the Nethybridge meeting in 1975. In 1976, he arranged to revisit Meselson's laboratory for a month, where, armed with the methylation-defective (*dam*) mutant strain from Martin Marinus, he and Wagner planned and executed the first experiments with hemimethylated DNA containing mispaired bases and immediately observed a dramatic effect. These and other more refined experiments led to Wagner's Ph.D. thesis and to the paper already mentioned, in which Meselson formally speculated about the role of DNA methylation as the basis for strand discrimination during mismatch excision repair in *E. coli*. "Matt really wasn't very keen on us doing those experiments," recalled Radman. "Maybe he thought that using the *dam* mutant was too risky. But once we had the first results he immediately recognized their significance." Some of these historical events are documented in a short paper published by Radman, Wagner, Barry Glickman (who as recounted below, together with Radman, isolated mutants defective in the *E. coli* mismatch correction enzyme machinery), and Matt Meselson. This paper constitutes part of the published proceedings of a meeting largely devoted to DNA repair which was organized and convened by Marija Alecevic in the magnificent coastal town of Tucepi, in the former Yugoslavia, in 1979.

> This paper is a chronological review of largely unpublished experiments started in 1976 in the Harvard Biological Laboratories when M. R. was a visiting scientist. . . . The idea that [*E. coli*] mismatch repair could function as an error correction system in DNA replication, using the transitory under-methylation of newly synthesized DNA strands to discriminate between old and new strands, was proposed by one of us (M. M.) and was discussed during the 1975 EMBO Recombination Workshop in Nethybridge, Scotland. This idea was consistent with the observations that . . . *E. coli dam* mutants, deficient in the major adenine methylation [three papers published by Martin Marinus and his coworkers between 1975 and 1979 are cited here] showed spontaneous mutator effects.

Once the potential role of strand methylation in mismatch repair was unequivocally identified, Meselson and his colleagues got wind of the fact that Paul Modrich and his student Gail Herman were investigating the biochemistry of DNA modification and restriction in great detail and knew how to methylate DNA in vitro. "So, I called Paul and asked him whether we could prepare phage λ DNA with different genetic markers, ship it down to him, and have him methylate it properly," recalled Meselson. "Then we would separate the strands and make various hybrid molecules to determine the repair patterns. So that's what we did and we showed conclusively that only hemimethylated DNA was properly repaired." Meanwhile, upon returning to his laboratory in Brussels, Radman began to pursue the genetic requirements for mismatch repair in *E. coli*, and together with Barry Glickman, now at the University of Victoria in Canada, found that a series of *E. coli* mutants called *mut* were defective in mismatch repair. These mutants later constituted an essential genetic framework for the elegant biochemical studies of Modrich and his colleagues that culminated in the purification and characterization of all the proteins involved in mismatch correction in *E. coli* and in the reconstitution of this system in vitro.

Paul Modrich, a professor at Duke University, independently evolved an interest in the mechanism of mismatch repair at the time that he was a graduate student with Robert Lehman at Stanford, which was where I first had the pleasure of getting to know him. "It was sort of an esoteric interest of mine for a long while," Modrich recounted. "But we really got moving on this problem as an extension of our work on the biochemistry of restriction and modification. This work gave me the idea of using restriction recognition sites as substrates for recognizing mismatches." (How curious a coincidence that both Meselson's and Modrich's interest in mismatch correction stemmed from their entirely independent studies on modification and restriction of DNA.) He laughingly recalled that, "for years I couldn't get anyone in the lab to work on this problem. I guess it took until about 1979 or 1980 before we first tried this assay." In this ingenious assay, Modrich and his colleagues engineered mismatches in known restriction sites in DNA molecules, which precluded their cleavage by the cognate restriction enzyme. Repair of the mismatches in *E. coli* extracts restored the restriction sites, which could then be recognized. Utilizing this assay, Modrich's laboratory went on, as just mentioned, to decipher essentially the complete biochemistry of mismatch repair in *E. coli*.

CHAPTER 6

The Remarkable SOS Phenomenon

SOON AFTER THE DISCOVERY of excision repair by Dick Setlow and then by Paul Howard-Flanders in the mid nineteen-sixties (see Chapter 3), Ruth Hill obtained a strain of *E. coli* called WP2 from Evelyn Witkin and isolated a UV-sensitive derivative that she designated WP2s. This strain, distinct from Hill's earlier and more extensively studied *E. coli* B_{s-1} mutant, also proved to be defective in excision repair. She observed that the WP2s strain sustained significantly more mutations in its genome than the parent excision-proficient strain following exposure to low doses of UV light. In her retrospective review entitled "Ultraviolet Mutagenesis and the SOS Response in *Escherichia coli*: A Personal Perspective," Evelyn Witkin wrote: "I confirmed and extended her [Ruth's] observations as did others ..., and drew the conclusion that excision repair is a relatively error-free process, whereas unexcised UV photoproducts are far more likely to cause mutations." The central issue of UV-radiation-induced mutagenesis circa the mid nineteen-sixties could now be succinctly stated: "It became reasonable to think of UV mutagenesis as error-prone replication across an unexcised pyrimidine dimer or other UV lesions in the template."

To this day, the precise nature of this "error-prone replication" is not fully understood in detailed biochemical terms. However, we know that the replication of DNA across a pyrimidine dimer is a complicated biochemical process requiring the regulated expression of a number of genes that are members of a large network. The members of this network are collectively regulated by a genetic pathway called the SOS pathway, a colloquialism whose origin I will explain shortly. Hence, this genetic network is sometimes referred to as the *SOS regulon* or *SOS system*. The transcription of SOS genes is upregulated when bacterial cells are exposed to various types of DNA damage, including UV radiation. Indeed, UV radiation is one of the most potent inducers of the SOS system. Here is how the basic elements of this regulation work.

SOS genes are normally repressed by the binding of a protein called LexA. The induction (upregulation) of these genes is negatively controlled, that is, derepression requires removal of the bound LexA protein from the operator site to which the protein binds in the repressed state. This derepression is effected by proteolysis of bound LexA protein through an autocatalytic mechanism which involves the participation of an essential coprotease called RecA protein. LexA and RecA proteins are therefore central players in this repression-induction scenario. Derepression of SOS genes can be activated by the exposure of cells to UV radiation or to many other types of DNA damage and by a variety of physiological perturbations that do not involve the direct exposure of cells to such damage. It is believed that the signal that is shared by all of these states, and that activates the RecA and LexA proteins to catalyze the destruction of bound LexA protein, is the exposure of regions of single-stranded DNA in the genome. If cells are subjected to UV radiation, such single-stranded regions arise when they attempt to replicate DNA containing UV photoproducts.

When bacteriophage λ correctly integrates its genome into a particular site in the bacterial chromosome, it is referred to as being in the prophage state. The maintenance of this state is also effected by negative gene regulation. But, in this case, repression is effected by a protein called the phage λ repressor. Interestingly, λ repressor protein is also subject to destruction by the RecA coprotease when cells are exposed to DNA damage, an event that results in replication of the phage genome, the formation of multiple new phage particles, and ultimately in lysis of the bacterial host cells. This series of events is termed phage λ induction. As you will see, the participation of RecA protein in the induction of both the SOS phenomenon and the formation of phage λ from the prophage state proved to be central to elucidating the fundamental nature of the SOS response. The discovery of this regulatory system and the essential elements of its operation represents a singular milestone in the history of our comprehension about how living cells respond to DNA damage, and not so incidentally, how genes are regulated in bacteria.

This chapter embraces the intellectual and experimental contributions of a large number of investigators among whom were two principal players, Evelyn Witkin, whose early history as a scientist we have already encountered (see Chapter 3), and Miroslav Radman, now a research director in the National Center for Scientific Research (CNRS) at the Jacques Monod Institute in Paris. You will recall from an earlier chapter that when Evelyn Witkin launched her career as a graduate student at Cold Spring Harbor Laboratory, she began studying UV-radiation-induced mutagenesis in *E. coli*. She observed that the strain she was working with, *E. coli* strain B, was prone to filamentation when exposed to UV light. This filamentation results from a failure of the cells to divide and separate as discrete entities following the replication of their DNA, due to a defect in septation. As a result, long chains of cells (filaments) or "snakes," as they were often called in laboratory jargon, formed. When I was a postdoctoral fellow in David Goldthwait's laboratory, Edward Kirby, a graduate student with Goldthwait, was working with a filamenting strain of *E. coli* for reasons that I will clarify later. I well recall Ed excitedly calling me over to his bench microscope to view these filamentous bacteria, whose length and snake-like appearance were truly quite impressive. Ed Kirby, who understood full well that his Ph.D. thesis was highly dependent on gaining some useful insights about how "these critters" formed, would stare at them for hours! By the mid nineteen-sixties, several investigators had begun to draw attention to similarities between UV-radiation-induced filamentation of *E. coli* strain B and λ induction by UV light in *E. coli* K12 strains carrying the phage in the integrated (prophage) state. Let us first briefly examine some of the early history of the discovery of λ induction.

As recounted by Horace Judson, the phenomenon of prophage induction has its historical roots in the early nineteen-twenties, when scientists at the Pasteur Institute in Paris noted that:

> A phage that would multiply virulently in one strain of bacteria could sometimes be set to infect a second, closely related strain—and would vanish. It would not commandeer the bacterial biochemistry to make more of itself, would not lyse the hosts and burst forth. Yet something was in there, nonetheless. The bacteria themselves would multiply unimpeded. Most peculiarly, they were immune to any further infusions of the same phage into their culture.

This state, called *lysogeny*, represents a dormant phase in the phage life cycle during which its genome is stably integrated and maintained in that of the host through multiple rounds of bacterial DNA replication. Judson wrote:

> But soon, free virus would be found in the culture, detectable in the usual way . . . [Max] Delbrück [who devoted much of his life to the study of the lytic phase of the phage life cycle] flatly refused to believe in lysogeny; a bacterial culture with a few persisting phage particles contradicted his fundamental proof that a bacterium infected by phage bursts, minutes later, and releases about a hundred new particles.

According to Ernst Peter Fischer and Carol Lipson, "the reason for [Delbrück's] resistance was the phage treaty [see Chapter 1], confining work to a common group of [T-type] phages. All bacterial viruses included in the phage treaty were virulent; with them, lysogeny did not occur [according to Delbrück!]." The disappearance (lysogeny) and reappearance (induction) of phage was poorly understood for many years, but was systematically investigated by the noted French microbiologist, André Lwoff, who in 1949 published the correct hypothesis that during the lysogenic state the infecting phage had somehow "merged into the chromosome of the bacterium," a form of the phage that he called prophage. On the basis of the profound variability of prophage induction in the laboratory, he further posited that induction was related to environmental changes. So, he set about trying to identify these changes. Judson abstracted Lwoff's accounting of the experiments that finally led to reproducible induction by exposing lysogenic cultures of *E. coli* to UV radiation. This accounting was appropriately included in the Festschrift prepared for the celebration of Max Delbrück's 60th birthday in 1966.

> Our aim was to persuade the totality of the bacterial population to produce bacteriophage. All our attempts—a large number of attempts it was—were without result . . . Yet I had decided that extrinsic factors must induce the formation of bacteriophage . . . Our experiments consisted in inoculating exponentially growing bacteria into a given medium and following bacterial growth by measuring optical density [that is, the turbidity of the bacterial mist, even though the particles were not visible individually]. Samples were taken every fifteen minutes, and the technicians reported the results. They (the technicians, that is) were so involved that they had identified themselves with the bacteria, or with the growth curves, and they used to say, for example: "I am exponential," or "I am slightly flattened." Technicians and bacteria were consubstantial.
>
> So negative experiments piled up, until after months and months of despair, it was decided to irradiate the bacteria with ultra-violet light. This was not rational at

all, for ultra-violet radiations kill bacteria and bacteriophages, and on a strictly logical basis the idea still looks illogical in retrospect. Anyhow, a suspension of lysogenic bacilli was put under the UV lamp for a few seconds.

The Service de Physiologie Microbienne is located in an attic, just under the roof of the Pasteur Institute, with no proper insulation. The thermometer sometimes rises in a manner that leaves no conclusion other than that the temperature is high. After irradiation, I collapsed in an armchair, in sweat, despair, and hope. Fifteen minutes later, Evelyne Ritz, my technician, entered the room and said: "Sir, I am growing normally." After another quarter of an hour, she came again and reported simply that she was normal. After fifteen more minutes, she was still growing. It was very hot and more desperate than ever. Now sixty minutes had elapsed since irradiation; Evelyne entered the room again and said very quietly, in her soft voice: "Sir, I am entirely lysed." So she was: and the bacteria had disappeared! As far as I can remember, this was the greatest thrill—molecular thrill—of my scientific career.

The unraveling of the mechanism of prophage induction is one of the great chapters in the history of molecular biology and in French science. It represents one of the two primary cornerstones in the history of gene regulation, the other being the regulation of the enzyme β-galactosidase in *E. coli*. Incredibly, both discoveries transpired contemporaneously at the Pasteur Institute in Paris; one in the laboratory of André Lwoff, the other in the laboratory of Jacques Monod. François Jacob, Monod's principal coinvestigator in elucidating the nature and regulation of the *lac* operon in *E. coli*, related to Judson that when (after much pleading) he was finally accepted to work with Lwoff at the Pasteur Institute in 1950, he found that:

> The lab was divided in two parts. There was a corridor between them; at one end, Lwoff was pouring UV light on lysogenic bacteria, including phage, and at the other end of the corridor Monod was pouring galactoside derivatives on *coli*—derived from the intestines of André Lwoff—to get enzyme induction. And the great joke was about induction, because each was "inducing" according to his own fashion, convinced that the two phenomena had nothing in common except the name.

In time, this extraordinary coincidence was happily recognized in full measure, and by the end of the decade the essential mechanism of the induction of β-galactosidase and of λ prophage were understood in principle, and were appreciated as parallel examples of negative gene regulation. Both were correctly reasoned to involve the derepression of the (repressed) expression of target genes (*operons*) by the binding of a specific repressor molecule to regulatory (*operator*) regions in the promoter of the target genes. By 1966, both the Lac and λ repressors were isolated by Walter Gilbert and Mark Ptashne respectively, both at Harvard University, and were shown to be proteins, as predicted by the French investigators. It was assumed that UV radiation and other exogenous factors known to induce phage λ inactivated the phage repressor. The precise mechanism of this inactivation was unknown, but by 1967 it was established that UV light did not inactivate λ repressor protein directly. This inactivation was believed with good reason to involve a series of undefined biochemical events which required postirradiation protein synthesis. The phenomenal contributions of Lwoff, Monod, and Jacob were recognized by their collective award of the Nobel Prize in Physiology and Medicine in 1965.

The seminal observation that induction of phage λ was initiated by UV radiation and other DNA-damaging agents was of more than passing interest to some students of DNA damage and its physiological consequences. In a paper published in 1967 entitled "The Radiation Sensitivity of *Escherichia coli* B: A Hypothesis Relating Filament Formation and Prophage Induction," Evelyn Witkin, who had never fully abandoned her interest in the phenomenon of filamentation of UV-irradiated *E. coli* that she observed years earlier, elaborated the phenomenological similarities between filament formation in UV-irradiated *E. coli* B and prophage λ induction in *E. coli* K12. Borrowing from the recently established molecular biology of prophage induction, she hypothesized that UV-induced filamentation worked in essentially the same way. She suggested that:

> (1) Strain B contains a repressor which, like the repressor of bacteriophage lambda, is inactivated by a complex process that starts with the presence of replication-blocking lesions, such as pyrimidine dimers, in the DNA. (2) This repressor (repressor B) is inactivated after UV only if protein synthesis occurs before the repair of pyrimidine dimers in the DNA is accomplished. (3) The inactivation of repressor B induces an operon (operon B), which may be part of an integrated episome or part of the bacterial genome proper. A product of operon B (presumably a protein) is an inhibitor of cell division or can lead indirectly to the production of such an inhibitor.

When I prodded Witkin about the origins of this hypothesis in her conscious thinking, she paused for a long time and replied:

> Gosh, I'm not sure I can answer that accurately. I had always thought of filamentous growth as behaving as if there were an inducible inhibitor of replication in the cell. My Ph.D. thesis basically worked out that the sensitivity of *E. coli* B to UV radiation was due to this filamentous growth. I had done a lot with the strain that didn't get into my papers but that gave me the feeling that what was happening was the induction of an inhibitor. For instance, I was aware of the liquid holding experiments of Aldous and Roberts. I knew that these were done with *E. coli* B and that the resistant strain B/r did not show liquid holding recovery. But liquid holding meant leaving the cells in saline—in a nongrowth medium. It required growth medium to show that the cells underwent filamentation. These were things that I didn't write about or put into words, but it was in the back of my mind that the cell needed to have active protein synthesis in order to die from filamentous growth. Then I guess just reading about what was being learned about λ induction—it just snapped together somehow.

In her 1967 paper, Witkin cited work by Israel Hertman and Salvador Luria in which it was shown that the induction of phage λ was dependent on an *E. coli* gene called *recA*, the discovery of which I will narrate presently. As already indicated, *recA* is a central player in the induction of both phage λ and the entire panoply of SOS genes. But, at that time, it was not known that filamentation (which is in fact part of the SOS phenotypic repertoire) was also dependent on the *recA* gene. In citing the work of Hertman and Luria, Witkin simply wished to stress the parallel complexity of phage induction and the induction of filamentation. At that time, the role of the *recA* gene in the former process did not strike her as relevant to the latter process. Just months after Witkin's paper was published, the merit of her hypothesis was strongly reinforced by the publication of work from David Goldthwait's laboratory, where, as I indicated, I had a first-hand voyeuristic introduction to filamentation in bacteria. In 1963, Goldthwait spent a sabbatical leave at the Pasteur Institute with François Jacob. Aside from his considerable scientific fame, the fact that Jacob worked in Paris did not encumber the attraction of his laboratory to American scholars seeking a place to enjoy a year away from home. Alan Campbell (who subsequently made substantial contributions to the biology of bacteriophage λ, especially the lysogenic state) worked as a postdoctoral fellow with Jacob, and during his tenure he isolated a phage mutant that was noninducible following exposure of lysogenic strains to UV radiation. This strain provided a powerful approach for dissecting the genetic requirements for induction. Goldthwait worked with a novel thermoinducible strain of *E. coli* which had been isolated in Jacob's laboratory, and which carried a thermosensitive mutation in the *bacterial* rather than the phage genome. At 30°C, the lysogenic strain grew normally, but at 40°C, phage production was induced. Goldthwait brought this mutant (called T-44) back to his laboratory at Case Western Reserve University, where it was cured of prophage and given to Edward Kirby to characterize as his Ph.D. thesis project. Ed was afflicted with hemophilia and his adventures with T-44 were for him a prelude to a career in biochemistry which he has devoted to the study of blood clotting. He was a delightful and extraordinarily bright young man who made a lasting impression on me and who provided me with a solid introduction to phage genetics and biology.

Among the phenotypes that Ed uncovered in T-44 was the observation that at elevated temperatures the strain underwent filamentation, and that the conditions for filament formation were very similar to those for thermoinduction of the prophage. This led to the reasonable conclusion that a mutation in a single gene probably determined UV-induced filamentation and induction of phage λ. But, Kirby and Goldthwait did not pursue this gene and they failed to recognize that they were sitting on a genetic gold mine. Their strain was in fact carrying a temperature-sensitive mutation in the *recA* gene, the gene that encodes the coprotease required for both λ induction and induction of the SOS genes, one of the phenotypic consequences of which, as just mentioned, is filamentation.

Meanwhile, other provocative biological relationships to λ induction were emerging. Studies by Matt Meselson and others were beginning to shed some light on the process of genetic recombination in bacteria and supported the notion that recombination involved the enzyme-catalyzed breakage and rejoining of duplex DNA molecules (see Chapter 5). At this time too, a picture of the genetic complexity of recombination was beginning to emerge due largely to the efforts of Alvin (John) Clark and his coworkers. Clark joined the faculty of the Department of Microbiology at the University of California, Berkeley in 1962. In his first semester, he was teaching a bacterial genetics course to a class of graduate students that included Ann Dee Margulies. Ann had graduated from Radcliffe and had learned something about recombination from Matt Meselson at Harvard. She was intrigued with this topic and one day she marched into Clark's office and challenged him with the leading question, "How does recombination work?" Clark admitted that he knew very little about the subject, but in appropriate professorial style he engaged his student in an extended Socratic dialogue and extracted from her a feasible experimental game plan for attacking this problem, beginning with the isolation of recombination-defective mutants. Margulies enthusiastically immersed herself in this project in Clark's laboratory. But, ironically it turned out that the instigator of Clark's life work had a greater interest in the practice of recombination between humans than its mechanism in *E. coli*! According to Clark:

> She was most interested in getting married. So, during her tenure in my laboratory, she would spend evenings entertaining potential suitors; baking pies for them and things like that. I did most of the work myself because she was always baking pies. In the end, she did not want a professional career as a scientist. She settled for a Master's degree, got married, and had children.

Clark and Margulies established an elegant screening protocol which allowed them to identify mutant colonies that failed to undergo a specialized form of conjugal recombination. Two mutants so identified were characterized further and shown to be defective in general recombination. Additionally, these strains were found to be quite sensitive to UV radiation. This phenotype was deliberately sought because Clark and Margulies were aware of the fact that soon after their studies on excision repair, Dick Boyce and Paul Howard-Flanders had discovered an excision-*independent* mechanism whereby *E. coli* cells can tolerate the lethal effects of UV radiation. This process was shown to involve recombinational events and was called *postreplicational recombinational repair*, or simply *postreplication repair*. Howard-Flanders and his collaborators predicted that mutant strains of *E. coli* should exist which were defective both in recombination and hypersensitive to UV radiation, a prediction fulfilled by the mutants isolated by Clark and Margulies. Clark called his recombination-defective mutant strains *rec*, and later one these was designated *recA*.

In 1990, the *E. coli* RecA protein had attained a level of interest that prompted an international conference in Saclay, France. In her introductory lecture at the conference, Evelyn Witkin remarked that:

> Every generation of biologists lives through thrilling discoveries that instantly transform their view of life, like sudden shifts in the pattern of a kaleidoscope. To those who come later, reading about the same advances in textbooks may evoke admiration and intellectual pleasure, but not the intense joy that they, in turn, will feel when they open a journal, or sit in the audience at a conference, and unexpectedly find that one of life's dark secrets has suddenly been flooded with light.
>
> I remember vividly one such moment of keen exaltation in 1965, when I came upon the paper by Clark and Margulies describing the first recombination-deficient mutant of *E. coli*. Genetic recombination is so much at the heart of biological diversity that the discovery of the gene encoding what appeared to be its master protein was truly exciting news...

In subsequent years, Clark and his colleagues deeply penetrated the alphabet with Rec⁻ mutant strains (at the time of writing, strains go at least as far as the letter R) as he systematically constructed a detailed genetic framework for recombination in *E. coli*. As a natural extension of the phenotypic characterization of the *recA* mutant, Clark and his coworkers examined recombination between bacteriophage genomes and between phage and host genomes. Recombination of phage T4 and of phage T7 was unaffected in the *recA* mutant. Katherine Brooks, another graduate student working with Clark, showed that the lysogenization of phage λ (a specialized form of recombination) was also normal. However, to her and her mentor's surprise, the reverse process, the UV induction of the lysogenized *recA* mutant, was defective. Following extensive experiments to eliminate alternative explanations, Brooks and Clark concluded in 1967 that:

> UV induction fails because immunity is not lifted in the irradiated lysogenic cells. This conclusion is further supported by the failure of superinfecting phage, sensitive to the repressor produced by the prophage, to multiply in the irradiated lysogenic host.

In their paper describing these findings, Brooks and Clark pointed out that several other investigators, notably Ogawa and Tomizawa in Japan, and Hertman and Luria at Massachusetts Institute of Technology, had independently observed the same phenomenon. (The latter is the work just alluded to in the context of Witkin's hypothesis of the inducible nature of filamentation.) Additionally, a few years prior to this publication by Brooks and Clark, Louis Siminovitch and his colleague Clarence Fuerst at the University of Toronto had searched for lysogens of *E. coli* that were defective in induction and found one (called T70) which they showed to be UV-sensitive and defective in recombination. So, a series of independent correlations were emerging. Witkin had correlated filamentation of UV-irradiated *E. coli* B with UV induction of phage λ and suggested that both functions involved gene derepression. The work of Kirby and Goldthwait indicated that both phenomena required a single function that was thermosensitive in their mutant called T-44. And at the same time, an independent correlation was emerging between UV induction of λ and a function required for genetic recombination.

In retrospect, it would have taken an extraordinary leap of intuitive genius for any single investigator to evolve the correct unifying hypothesis for these multiple independent observations. Before the waters get too muddied for the uninitiated reader, allow me to leap forward in history and inform you that RecA protein, the product of the *recA* gene formally discovered by John Clark and Ann Dee Margulies, is a most versatile protein. In addition to its role as a coprotease in the inactivation of the λ and LexA repressor proteins, it is also required for the process of genetic recombination, and this is the basis for the recombination-defective and recombinational repair-defective phenotypes of the *recA* mutant. (In fact, we now know that RecA protein has a *third* role specifically required for mutation induction.) But in the mid to late nineteen-sixties, too many pieces of the puzzle were missing to appreciate the dual functionality of RecA protein. Additionally, the role of *recA* in recombination was particularly confounding because several phenomena that in fact derive from the role of RecA protein in inactivating the LexA and phage λ repressors were quite reasonably (but incorrectly) rationalized in terms of this role. For example, in the letter in *Virology* in which they announced the isolation of the recombination-defective mutant T70, Fuerst and Siminovitch wrote: "These observations therefore lend support to the proposal by [Alan] Campbell that genetic crossover is involved in UV induction of prophage development." And in their paper published in 1967, Brooks and Clark commented that: "One can hypothesize that there might be a step in the lifting of immunity which is a recombination event or is biochemically similar to one of the steps in recombination."

Hence, even if Witkin had been aware of the dependence of λ induction on the *recA* gene at the time that she published her induction hypothesis in 1967, there is no compelling reason why she too would have not assumed that this reflected the role of the gene in recombination. Brooks and Clark actually went so far as to attempt to segregate genetic functions that specified recombination deficiency and loss of UV inducibility of prophage λ. They were unable to do so in the mere 40 recombinants they tested. Had they been unusually lucky, they would have succeeded because we now know that these phenotypes are indeed genetically separable. Indeed, mutants that were proficient in recombination but defective in the SOS response were isolated soon after by the French microbial geneticist, Raymond Devoret, whose intimate involvement with the history of the SOS story we shall encounter presently. "I found such mutants because I was able to *select* for them as noninducible lysogens after thymineless induction [the SOS response can be induced by starving cells for thymine] and testing them for recombination on replica plates," recalled Devoret. "Others, like John Clark, tried to isolate such mutants by using a brute force screening approach, which was very difficult to do."

But, being alleles of a single gene, such segregants are very rare and would have required much more extensive segregation analysis. John Clark was apparently unaware of Evelyn Witkin's hypothesis linking filamentation in *E. coli* B and prophage induction in *E. coli* K12 lysogenic for λ, and Witkin did not cite Clark's correlations between recombination deficiency and the λ noninducible phenotype. One group of investigators was primarily interested in genetic recombination. The other was transiently interested in filamentation, and as we shall see presently, more deeply interested in UV-induced mutagenesis.

One of the key missing pieces in the puzzle that, when finally put together, revealed the SOS phenomenon was the lack of experimental evidence for the existence of the "repressor molecule" hypothesized by Witkin, which, like the λ repressor, was inactivated by exposure of cells to UV radiation. But, as I just indicated, Witkin's interest in filamentation in *E. coli* B was transient. She did not deliberately set out to prove her repressor model. As recounted in an earlier chapter, her main interest was UV-induced mutagenesis. What by any standards of coincidence must be considered the most extraordinary twist of fate, her pursuit of this line of investigation unknowingly led her to the very mutant that held the clue to her repressor model—a mutant defective in the gene that encodes the LexA repressor. It turns out that mutagenesis in *E. coli*, which is induced by exposure of cells to UV radiation, like λ induction and filamentation, is yet another manifestation of the SOS response! But, it took a while for that proverbial coin to drop.

By the mid nineteen-sixties, Paul Howard-Flanders and his coworkers had isolated and genetically mapped an impressive armada of mutants of *E. coli* K12 that were either UV-sensitive, X-ray-sensitive, or recombination-defective, or manifested combinations of these phenotypes. A Dutch group from Rijswijk headed by Arthur Rörsch also isolated such mutants. Howard-Flanders called one of these genes *lex* (for *l*ocus for *ex*-ray sensitivity). (According to Raymond Devoret, the designation *lex* was originally coined by Howard-Flanders as a tribute to Lee Theriot [*lee* X ray], who actually isolated these mutants in his laboratory. The rationalization with "*l*ocus for *ex*-ray sensitivity" was post hoc.) The *lex* mutant was known to be sensitive to both X rays and UV radiation. By this time, it was generally appreciated that Ruth Hill's now extensively studied *E. coli* B_{s-1} strain was also abnormally sensitive to killing by both X rays and UV radiation. These phenotypes were determined by two distinct genetic loci called *hcr* (later *uvrB*) and *exr*. The *hcr* mutation was known to determine some of the UV-sensitive phenotype but not its additional sensitivity to X rays, while the *exr* mutation was known to cause increased sensitivity to both UV radiation and X rays. Hill's strain was also known to be nonmutable following exposure to UV radiation. This phenotype was especially interesting to Witkin, who "always had a lot of faith in what one can learn from revertants," and "wondered whether one could recover resistance to killing by UV radiation without affecting the nonmutable phenotype." So, she set about isolating and characterizing a large number of revertants of B_{s-1} with varying levels of UV sensitivity. "I was indeed able to isolate strains that were as resistant as the wild type and were still nonmutable by UV light," she recalled. Some of these had recovered their capacity for excision repair and were considered to be Hcr$^+$. Other revertants recovered the phenotype of UV mutability and these were considered to be Exr$^+$. In a paper published in the *Proceedings of the Brookhaven Symposia in Biology* in 1967, Witkin offered an hypothesis for the UV nonmutability of the *exr* mutants.

> Exr$^+$ strains must replicate their DNA after UV (or repair it by a mechanism not involving excision) in a way that frequently causes a change in base sequence in the neighborhood of an unexcised pyrimidine dimer . . . One proposed method that might be used by irradiated bacteria to replicate DNA containing pyrimidine dimers is to "bypass" the dimers during the first post-irradiation replication . . .

Aside from the fact that the essential features of this model of UV-induced mutagenesis turned out to be correct, perhaps its historically most intriguing aspect is that Witkin was unknowingly proposing a cardinal feature of the SOS response, a biological phenomenon that had already touched her life in the guise of the filamentous phenotype of the *E. coli* B strain while she was a graduate student (see Chapter 3). In 1967, she had no way of knowing that filamentation and UV-induced mutagenesis were mechanistically related. Once again, recombination was a confounding issue. Witkin noted that like the *recA* mutant, her UV nonmutable *exr* strains were defective in recombination, though the extent of the defect was not as marked as in the *recA* strain. In light of this observation, she was not especially surprised to discover that the recombination-defective *recA* mutant was also UV nonmutable. Indeed, by the late nineteen-sixties, the model of UV-induced mutagenesis by aberrant "bypass synthesis" had been adapted to incorporate the notion that the central feature of error-prone DNA replication transpired during DNA synthesis specifically associated with recombinational repair. In 1969, Witkin published a paper on the UV-mutability of recombination-defective strains of *E. coli* in which she documented that the *recA* mutant is not mutable by exposure to UV light. She wrote that:

> This result [that a *recA* mutant is UV nonmutable] supports the hypothesis that UV noninduced mutations in *E. coli* originate as errors in the recombinational repair of gaps. The *recA* gene product is required for *any* recombinational repair; the *exr* gene product is responsible for the inaccuracy of this repair.

It was subsequently shown that the *exr* mutation in Ruth Hill's strain was in fact an allele of the *lexA*$^+$ locus identified by Paul Howard-Flanders and his associates. It is this gene that specifies the LexA repressor protein which silences the SOS genes. Inactivation of both the LexA and λ repressors requires a functional *recA* gene. Witkin recalled that "later we mapped many of these mutants that acquired normal UV resistance but remained UV-nonmutable and they were a gold mine of *lexA* and *recA* alleles." But, as already mentioned, at that time Witkin was not in a position to exploit the potential clue that the inability of the *exr* (*lexA*) and *recA* mutants to be mutated by exposure to UV radiation might be connected by a relationship other than recombination. However, other provocative interpretations of the mechanism of UV-induced mutagenesis were beginning to circulate.

In her historical reflections on the emergence of the SOS phenomenon entitled "Ultraviolet Mutagenesis and the SOS Response in *Escherichia coli*: A Personal Perspective," published in 1989, Evelyn Witkin recalls that:

> In 1971, I received a memorandum from Miroslav Radman, who was then a postdoctoral fellow at Harvard. Radman had reinterpreted the "Weigle phenomenon," whereby UV-irradiated λ phage produce UV-induced mutations and exhibit high survival only if they infect host cells that have also been exposed to UV. He suggested that UV induces a mutagenic mode of DNA replication, which he called "SOS replication," and that both the Weigle effect and *bacterial* [my italics] UV mutagenesis are manifestations of this DNA damage-inducible activity, dependent on the *recA*$^+$ and *lexA*$^+$ gene products for its induction.

I have had the adventure (I know of no better word) of knowing Miroslav Radman for close to 30 years. We first met in 1969 at a meeting on DNA repair organized by Raymond Latarjet in Evian, France. A native of former Yugoslavia and now Croatia, Radman was then a postdoctoral fellow in Devoret's laboratory at the CNRS unit in Gif-sur-Yvette, a rural and charming suburb on the outskirts of Paris. Our acquaintance at that meeting was limited, but I was immediately impressed by the intelligence and wit of this amiable and energy-charged young man, an impression gained chiefly by sitting on the fringes of several animated outdoor evening discussions about Weigle reactivation and other related phenomena, which Radman and Devoret dominated. Since then, Miroslav and I have cultivated a firm intellectual association coupled with a close personal friendship, for we have identified much in common outside of our scientific interests, in particular a shared irreverence of many "traditional" attitudes. There is no one in the scientific community whom I know better and there are few whose intellect I respect more. Miro is without question one of the most scientifically provocative individuals I have ever met. His thoughts are never far from science and his mind is always teeming with hypotheses, a good number of which, like the SOS hypothesis, have turned up trumps. Planning a meeting of any substance on cellular responses to DNA damage without including Radman in the program is like contemplating a football homecoming party without the queen. Regardless of the topic in question, he can be relied upon to be stimulating, and as anyone who has ever chaired one of his formal talks can attest, he just as reliably will exceed his allotted time. But, if one is a member of the audience listening to a wound-up Radman and watching him furiously mopping his forehead as he battles with the clock, one often silently implores the session chair to give him the extra few minutes. His intellectual prowess, his rich sense of humor, and his enormous personal warmth and charm are exceeded only by his total disdain for the logistics of "normal" living. The word "deadline" is nonexistent in Radman's vocabulary, nor can he be counted on to show up in less than a few hours after most scheduled appointments. Hence, knowing Miroslav Radman is indeed an adventure!

When I asked Raymond Devoret, who as we shall see played a significant role in Radman's career development and has known him even longer than I have, whether he thought I was exaggerating Radman's scientific talents, he responded without hesitation. "No, not at all," he said. "But," he added quite seriously, "I wouldn't write about his lack of punctuality and so on. Because now Miro comes to the Academy wearing a suit . . . and he comes right on time! He's more mature." I could only smilingly reflect on the last time that I had scheduled any sort of appointment with Radman and the last time that I had seen him in a suit!

Radman received his undergraduate training at the University of Zagreb between 1962–1966. Here he had the good fortune of securing a part-time job assisting in the compilation of information for an ambitious Yugoslavian publication entitled *The Encyclopedia Moderna*. This work was intended to document the professional activities of leading intellectuals in Yugoslavia and was orchestrated by the famous Yugoslavian physicist, Evan Supek, a former student of Werner Heisenberg and Niels Bohr and a formidable intellect in his own right. Supek was quick to note the keen intellectual bent of the young Radman and steered him to Maria Drakulić, a biologist at the Ruder Bosković Institute in Zagreb, in order that she might direct Radman's undergraduate thesis, a requirement for all undergraduates at the university. Drakulić had studied with the noted Belgian biologist, Jean Brachet, and was an intimate friend of the late Ruth Hill, the discoverer of the UV-radiation-sensitive mutant *E. coli* B_{s-1}. "It was Professor Drakulić who initiated me into the intricacies of experimental biology and the many interesting problems of DNA repair," Radman told me. "One of the first journals I ever read was the special volume of *Radiation Research* published in 1966 that reported the proceedings of the DNA repair meeting in Chicago" (see Chapter 3). Radman was immediately fascinated by the work on excision repair reported by Setlow and Carrier and Boyce and Howard-Flanders. "I thought this was something really beautiful—to read about repair of damaged genetic information at the level of the DNA molecule," recalled Radman. And so an important convert was made to the field.

It soon became apparent to Maria Drakulić that Radman had exhausted the educational potential of the University of Zagreb. After he obtained his Master's degree in 1967, she urged him to attend the University of Brussels, which at that time was considered to be one of the major centers of molecular biology in Europe. Armed with an EMBO fellowship, Radman journeyed to Maurice Errera's laboratory to become a graduate student. He was given a free hand to study under the direct tutelage of any of the senior scientific staff in the laboratory and elected to work with Ann Roller, Matt Meselson's first graduate student at Caltech. His choice was predicated mainly on the fact that he (then) spoke very little French and Roller spoke German, a language that Radman was fluent in. By this accident of language, Radman was introduced to phage λ, an education that was rapidly intensified by his proximity to René Thomas's laboratory in the department and the presence of the Americans Peggy Lieb, Max Gottesman, and later Michael Yarmolinsky, as sabbatical visitors at the Pasteur Institute in Paris. "I fell in love with λ," Radman confided. "Every month a new gene was coming out of this system and I realized at once that I could use λ genetics and λ DNA to pursue my interests in DNA repair." Within two years, he had completed his Ph.D. thesis on aspects of excision repair and postreplication repair in phage λ. "I was very fortunate," Radman told me. "Close to my 25th birthday, I had five publications and my Ph.D. thesis completed."

In the late nineteen-sixties, Raymond Devoret, one of the leading bacterial geneticists in Europe, together with his Dutch colleague, Arthur Rörsch, mounted a series of laboratory courses for selected European students modeled very much along the lines of the summer Cold Spring Harbor Laboratory courses. Devoret informed me:

> We had money from EURATOM and we wanted to educate European students to modern techniques in DNA repair. When we planned the first course, Errera asked if we would take Miro, whom I had met when I visited Brussels, and so Miro attended the first course. I liked him from the first day and you could tell immediately that he was going to be the best student in the course. All the other students were running around not finding things, but Miro quietly examined the protocols we provided and immediately got started. He was very gifted technically.

Shortly after attending this course in 1969, Radman decided to pursue a year of postdoctoral study with Devoret and his senior colleague, Jacqueline George, before departing to the United States for a more extended postdoctoral fellowship with Matt Meselson that he had already arranged and acquired funding for. "Raymond was a leading bacterial geneticist," Radman told me. "I was trying to decide between his lab and the Pasteur Institute, but I opted for Gif because there was more DNA repair and mutagenesis work going on there."

During his final months in Brussels, Radman had developed an intense interest in a biological phenomenon discovered by Jean Weigle, a Swiss biologist at Caltech and a member of Delbrück's phage group there. In 1953, Weigle published the finding that now bears his name. He introduced this paper as follows:

> In the course of experiments designed for other purposes a paradoxical observation was made; phage λ, inactivated by UV irradiation, when adsorbed onto sensitive bacteria was reactivated when a further dose of UV was given to the phage-bacterium complexes.

This reactivation phenomenon, originally called UV-reactivation of λ by Weigle, is now referred to as Weigle or W reactivation, having been later graced with the name of its discoverer by Radman and Evelyn Witkin. An element of his observations that Weigle stressed in his 1953 paper but which is not evident from the term "reactivation" is that "among the reactivated phages a fairly large proportion were mutants." The significance of this observation was not lost to Weigle, who in fact entitled his paper "Induction of Mutations in a Bacterial Virus." Of course he had no sense of how increased survival and mutagenesis of UV-irradiated phage λ transpired. But he did not fail to stress that UV irradiation of *both* the phage and the host was necessary to observe these effects. In so doing, he sensitized future (attentive) readers to the crucial fact that UV radiation may be playing two different roles: providing a source of premutational damage in the phage genome and independently triggering a response that eventually resulted in mutations.

> Our experiments do not tell us what is the specific role played by the action of the UV on the phages and on the bacteria, respectively. They rather emphasize that both actions are necessary. This double causation is very obvious in the present case. It raises the question whether mutagenesis might not involve a similar double causation.

"When I defended my Ph.D. thesis," Radman recalled, "René Thomas, a member of my committee, really focused my attention on this. He asked me, 'Why do you think one needs to irradiate the host bacterium in order to observe mutagenesis of λ? Think about this Miro. There's something very interesting here.' I read Weigle's paper so many times that my copy was literally falling apart. Without this hint from Thomas, I don't think I would have come to the notion of the inducible nature of the SOS phenomenon."

While still in Maurice Errera's laboratory in Brussels, Radman began to actively pursue the mechanism of Weigle reactivation together with his colleagues, Martine Defais (whom you may recall from Chapter 4 was a member of John Cairns's group which elaborated the adaptive response to alkylation damage) and Perrine Fauquet. He maintained this interest after he moved to Devoret's laboratory. "I told Devoret that I wanted to understand this problem of why you had to irradiate *E. coli* in order to observe mutagenesis in the irradiated phage." At that time, Devoret and his Spanish colleague, Manuel Blanco, were testing a hypothesis developed by Weigle suggesting that irradiation of both λ and *E. coli* may be required to produce mutations by otherwise rare recombination events between similar sequences in the phage and bacterial genomes. But Radman wasn't persuaded. "I thought very early on that it's not recombination. I was for an inducible process turning something on, because we already had done experiments in Brussels suggesting this to me." These experiments showed that UV induction of lysogenic phage λ, induction of reactivation of UV-irradiated λ, and induction of mutagenesis of irradiated λ all transpired with identical kinetics. "I was struck by these parallel observations," Radman related. "I considered λ induction to be an established paradigm of an inducible phenomenon and I was convinced that the other phenomena were also the result of some inducible process." More importantly, the Brussels team had by then discovered a requirement for the *lexA* gene for W reactivation and also showed that this reactivation required active protein synthesis. A final death blow to the recombination hypothesis was the demonstration that the noninducible *lexA* mutant was perfectly competent for recombination and that known λ recombination functions were not required for the induction phenomena. When they published some of these observations in 1971, the authors stated that "it is worthwhile to notice that both *recA* and *exrA* (equivalent to *lex*) bacterial mutations suppress pleiotropically UV mutagenesis, UV reactivation, and prophage induction, suggesting that some common pathways might be involved in these phenomena."

In reflecting on his brief transition period in Devoret's laboratory between

Brussels and his more extended postdoctoral studies in the United States, Radman revealed that:

> The year was crucial for me because I had strong intellectual adversaries to challenge my ideas and I had a lot of time to think. When I arrived in Devoret's lab, they believed in the recombination model of Weigle mutagenesis. In the course of that year, they convinced themselves, and I believe that I helped convince them, that this was not the case. But I was constantly challenged to think about my alternative hypothesis. There was certainly no way you could fall asleep in Raymond's lab. During this time, I also absorbed an enormous amount of literature. I would go home in the evening with a lot of journals and books and I talked a lot with Devoret, Jacqueline George, Manuel Blanco, and others in the lab and my ideas were getting sharper and sharper. Before I left Europe for Matt's lab, my formulation of the basic SOS model was essentially complete and I presented it during a seminar that I gave to René Thomas's lab in the fall of 1970 just before I left for the United States.

Devoret concurred that Radman was an exciting and provocative presence in his laboratory at that time, but pointed out that he and Radman had philosophic differences. "I would disagree with his assertion that we firmly believed in Weigle's recombination model," he commented. "The main difference between the way that Miro and I operated then was that he wished to discard the model by the weight of pure argument, whereas I wished to disprove the model by the weight of data obtained by direct experimentation."

Another important influence of Devoret's laboratory stemmed from the fact that at the time that Radman joined that group, Devoret and his colleagues were investigating the phenomenon of indirect induction of phage λ. As Devoret described it:

> We had started investigating a phenomenon that I originally called the Borek-Ryan effect, in deference to the individuals who first observed it. These investigators irradiated a nonlysogenic strain of *E. coli* and mixed the bacteria with unirradiated bacteria that were carrying λ as a prophage, and they observed lysogenic induction of the phage. They suggested that the irradiated cells released some sort of chemical that got into the lysogenic strain and caused induction. We showed that the irradiated strain had to be male and the unirradiated lysogenic strain had to be female, and that the irradiated F-factor DNA in the male was being transferred to the female strain and this is what caused λ induction. Jacqueline George and I discussed these experiments a lot with Miro and he immediately recognized that we had a system in which one could separate the lesions in the DNA from the outcome of the lesions.

"I recall these experiments very well," confirmed Radman, "and I suggested to Raymond and Jacqueline that we should ask whether one can also observe W reactivation and W mutagenesis of λ by indirect induction." The experiments worked beautifully and provided another striking parallel between λ induction and the Weigle phenomenon. "We started these experiments in 1970, but Jacqueline was a very careful experimentalist and the results weren't submitted for publication until mid 1973." By this time, the SOS hypothesis had matured to the point that the authors could confidently suggest that "the occurrence of indirect UV-reactivation provides evidence for the existence in *E. coli* of an inducible error-prone DNA repair mechanism."

At that time too, Devoret's laboratory showed a requirement for the *recA* gene for indirect induction of λ. Devoret told me:

> We knew about the experiments of Hertman and Luria on *recA*. But I decided to independently look for mutants that failed to show indirect induction. Sixty percent of the mutants were in the *recA* gene.

When I asked Devoret whether he was aware of the T-44 mutant that Goldthwait had isolated at about that time, he responded:

> Yes, we were. But this was pursued by Jacqueline George after she moved to the Jacques Monod Institute to work with Gerard Buttin. Jacqueline was a staunch communist who had been very involved in the cultural revolution in France in 1968. She was bitterly disappointed and embarrassed by its failure and told me that she simply couldn't face having to acknowledge this failure to all the people she knew in the institute at Gif. So she moved to the Jacques Monod Institute where she, Castellazzi, and Buttin showed that T-44 was a *recA* allele.

This group also reported in their paper published in 1972 that W reactivation and W mutagenesis can result by simply shifting the temperature-sensitive T-44 mutant to a higher temperature, an experiment that Radman recalls discussing in detail with Jacqueline George.

> I suggested to her that we should look for yet another parallel between λ induction and W reactivation by showing that just as the temperature-sensitive mutant *dnaB* induces λ when DNA replication is blocked, W reactivation and W mutagenesis may be induced when T-44 is shifted to an elevated temperature. I was also interested in this experiment because I thought it would show that the induction wasn't really related to the presence of any specific lesion in the DNA, but that it received a signal from the stalled replication fork. So, before I left for Matt's lab in October of 1970, all these experiments were underway. When I was at Harvard, Jacqueline kept me informed of the results as they emerged.

What did Devoret think of Radman's hypothesis to explain W reactivation and W mutagenesis? "He was right of course," Devoret told me in his delightful French accent. "But at the time the experimental data simply weren't there. Besides which, I was deeply immersed in my own interests; the problem of lysogenic induction and its many experimental intricacies, and his ideas were too much of a jump for me to consider seriously. Not that I didn't share his idea—but I was still skeptical at that time." In retrospect, Radman agrees that his association with the Devoret laboratory, brief as it was, was crucial to his evolution of the SOS hypothesis and the notion of a third DNA repair pathway that was distinct from excision repair and from recombinational repair. Apparently, the French Academy of Sciences shared this view of history because in 1992, Radman and Devoret were corecipients of the Charles Leopold Meyer Prize, a prize enjoyed in earlier years by Watson and Crick, Jacob and Monod, and also by Evelyn Witkin, Mark Ptashne, and Walter Gilbert.

In reflecting about the time of his move to Harvard, Radman ruefully recounted that:

> I had to abandon my full-time preoccupation with SOS just when it was all getting to be most exciting. However, I wanted to get out of radiobiology for a while and I wanted to work on recombination. Recombination was my new dream.

But, in fact, Radman did not immediately stop thinking or talking about his idea of inducible mutagenesis at all. He is very fond of conceptualizing his ideas in cartoon form and those who have heard him talk at seminars or meetings will recall that he rarely uses slides in his presentations, preferring instead to use hand-drawn "overhead" representations of what he wants to communicate. Those who know him especially well will also know that most of these drawings are put together in the wee hours of the morning of his scheduled talk. At more than one meeting, I have left Miroslav at his hotel room door well after midnight and well into our cups, and heard him say, "Well, I'd better get this talk together." Soon after arriving in Cambridge, Massachusetts, but before he immersed himself in his new interest in recombination, Radman was up late one night cartooning his SOS model. He excitedly related to me:

> I drew this *E. coli* cell with a sort of a compartment in which I put the viral genome showing it undergoing mutagenesis. And then quite suddenly it struck me that why on earth would there be a special "compartment" just to mutagenize viruses. Surely this was a strategy that the host had evolved for its own genome. I was so excited when I had that insight that I immediately began to write this memo. I did not sleep a wink that night. The following morning my fiancée [later Radman's first wife] typed this for me on a typewriter borrowed from Matt's lab.

In a later refinement of this document, which Radman entitled "SOS Replication: A Distinct DNA Replication Mechanism Which Is Induced By DNA-Damaging Agents," a copy of which I obtained from Evelyn Witkin who told me that "this was one of those things that I thought I should save because somehow I could sense its importance," Radman wrote:

> Weigle phenomenon: a general inducible replication repair process in *E. coli*? . . . . Weigle's phenomenon (as discovered and defined by Weigle in 1953) is a host-dependent UV induced increase in survival of infecting UV damaged phage. It is a very efficient repair process. A simple question can be asked: does this repair process operate also on bacterial (host) DNA? . . . . I will try to deduce the existence of the Weigle phenomenon in the uninfected *E. coli* cells from the known discrepancies between the lethality of pyrimidine dimers in bacterial DNA and in λ DNA.

Radman then proceeded to document evidence that the number of pyrimidine dimers required to constitute a lethal hit in *E. coli* and in phage λ were very different, but this difference could be accommodated if one assumed that in addition to excision repair and postreplicative recombinational repair, a third repair process exists in *E. coli*; namely SOS repair.

> What kind of repair reflects Weigle effect? I would like to suggest that it is a replication repair: "SOS-replication." By replicating damaged DNA by the SOS-replication, some incorrect bases are necessarily expected to be built in the new chain opposite Py-Py dimer, which would account for (1) the majority of UV induced mutations in *E. coli* being single base substitutions.

Where did the term SOS come from? Radman smiled:

> Well, you know that I come from an island, and my father [a fisherman on the island of Hvar] lives on the sea. SOS is the international distress signal to save endangered life on the sea. I viewed error-prone DNA replication as an emergency response of the cell to DNA damage, whereby the viability of the cell could be restored at the cost of accumulating mutations.

During the period immediately prior to departing for the United States and his fellowship with Meselson (an association that as recounted in the previous chapter contributed little to our knowledge of recombination, but did initiate Radman's contributions to the elucidation of mismatch repair), Radman was unaware of Witkin's hypothesis published in 1967 suggesting a linkage between filamentous growth and prophage induction. Similarly, when she presented a seminar at Harvard in December 1971, Witkin was unaware of the work of the Belgian group. Radman had met Evelyn Witkin rather briefly some years earlier when he was a student, but was fully aware of her considerable reputation in the field of UV-induced mutagenesis. Upon hearing about her impending visit to Harvard, he wrote to her in late November 1971.

> Dear Dr. Witkin,
>
> I just saw the announcement of your seminar to be given on December 2 in Boston. I shall be very pleased to see you and hear you again.
>
> Since October 1970 I have been working in Matt Meselson's lab and have essentially quit radiobiology. I have detected an endonuclease which is specific for mismatched base pairs in DNA and thus is a good candidate as an enzyme involved in gene conversion of mismatch correction type.
>
> But as you can see from the enclosed paper I cannot stop to think about radiobiology. I would be very thankful to you if you could read this typewritten text (which is just a discussion letter I have sent to my friends who are interested in repair and mutagenesis) and tell me your objections. I hope you will have a few minutes after your seminar in Boston. Although I wrote this text one year ago it still seems attractive to me that UV-reactivation or "SOS replication" is a third repair system (inducible) in E. coli and must be distinguished from REC repair.... I wish I could exchange some of these ideas with you.
>
> Yours sincerely.
>
> Miroslav Radman

Witkin told me:

> I had met Miro a couple years earlier when Maurice Errera brought him to my lab in Brooklyn for a visit. Needless to say, I had been impressed. I read Miro's memo with interest, but, although I found it convincing for Weigle reactivation and [Weigle] mutagenesis, I did not think there was any direct evidence for inducibility of *bacterial* [my italics] UV mutagenesis, other than the suggestive *rec/lex* genetic requirement. Miro had incorrectly cited a requirement for post-UV protein synthesis as evidence for inducibility of the activity responsible for bacterial UV mutagenesis. He may have been thinking of MFD [mutation frequency decline, see Chapter 3], but this was unique to suppressor mutations. There was actually no such evidence, as protein synthesis had eventually to be allowed for expression and scoring of mutants, and therefore one could not know if it was also required for the mutagenic event itself. When I met Miro at Harvard in December 1971, I pointed this out to him. He was a bit disappointed that I wasn't altogether convinced. He also mentioned that none of the others to whom he had sent the memorandum [including Walter Gilbert, Mark Ptashne, René Thomas, Raymond Devoret, Paul Howard-Flanders, and Matt Meselson] had responded, even to be critical.

Over numerous glasses of slivovitz (or related beverages of similar alcoholic intensity) dispatched in subsequent years, Radman sometimes confided to me his disappointment at the lack of responsiveness to his memorandum in which he outlined the SOS model. This disappointment was perhaps primarily related to the fact that the scientific community to which he now wished to belong, the molecular biologists other than radiobiologists, whom he too believed were "not taken very seriously," weren't especially interested in his ideas. When we rehashed this in the fall of 1995, he admitted that in retrospect:

> I was trying to convey too much theoretical information that was not yet substantiated by experimental evidence. I now realize that this would have given indigestion and constipation to any sane person. And the brighter the person, the faster they became bored. I also now realize that scientific notables at Harvard such as Matt Meselson, Walter Gilbert, and Mark Ptashne were too busy to be seriously bothered with this effusive young postdoc from Yugoslavia and his ideas on the mechanism of mutagenesis caused by UV radiation. I remember that very soon after I arrived at Harvard, Wally Gilbert and his wife Celia generously invited Birgit [Radman's then fiancée] and me for lunch at his home one Sunday. He was extremely open-minded and more than willing to listen to the story. But it was too much. Wally told me that he could readily understand that if one directly damaged the operator of an inducible gene by UV radiation, one might expect some sort of regulatory upset. But he couldn't buy the notion of the cell somehow sensing randomly distributed damage in the genome and setting in motion a complex genetic program that turned on so many functionally unrelated genes in such a coordinated manner. Part of the problem in selling this notion was that SOS was not only about mutagenesis. It was also a new paradigm for complex gene control.

During their meeting in late 1971, Witkin informed Radman of her earlier suggestion that filamentous growth in *E. coli* and prophage induction were similarly regulated. But, at that time, she was unaware of the *recA* dependence of filamentous growth, and its *lexA* dependence was not yet known. So filamentation "was not an obvious candidate for *recA/lexA* control." In short, there was not enough evidence to persuade her of Radman's view that UV-induced mutagenesis in *E. coli* was an inducible phenomenon mechanistically linked to W reactivation, W mutagenesis, and prophage induction.

In September of 1972, Witkin was preparing a talk for a conference to be held in Bethesda on "The Genetic Control of Mutation." In this address, she planned to discuss recent confounding observations on UV-mutability in John Cairns's *polA* mutant. "I had just received the two reprints of Castellazzi et al.'s 1972 papers on the *tif* (T-44) mutant," she told me. As already indicated, this mutant of the thermoinducible strain first reported by Goldthwait and Jacob and originally designated T-44 mimicked several UV-inducible responses such as filamentation and λ induction upon simple exposure to elevated temperature. (Parenthetically, the genetic designation *tif* stands for *t*hermo*i*nducible *f*ilamentation. Devoret pointed out to me that in French, the word "tif" is French slang for a piece of hair, appropriately descriptive for the filamentous state of bacterial cells.) By 1972, as Witkin described it:

> Castellazzi and his coworkers were able to elegantly explore the linked regulation of prophage induction and filamentous growth, and showed that their expression required *recA* and *lexA*, making it appear likely that *tif* is an allele of *recA*. I had just read them with interest, and I went back to puzzling over my *pol* data, and trying to make sense of them. There was a specific moment which I shall never forget, when I had one of those marvelous eureka-like experiences and the pieces all fell together for me in a meaningful pattern. I quite suddenly knew, with certainty, that Radman was right about bacterial UV mutagenesis depending on an inducible function.

What Witkin had observed in the Cairns mutant was that whereas at higher doses of UV radiation mutation frequency was normal, at very low doses the strain was highly mutable. The "eureka moment" was her intuitive flash that this hypermutability might reflect *hyperinducibility* due to the abnormally slow closing of single-stranded gaps during excision repair in the polymerase mutant.

> I excitedly looked up the published dose curves for the *pol* mutant's induction of λ prophage and for W reactivation and found that indeed the *pol* mutant induced these functions in the same low UV fluence range. I was now absolutely convinced that Radman's SOS replication was the mutagenic activity needed for bacterial UV mutagenesis—probably as Defais et al. and Radman had suggested, the same activity that caused Weigle reactivation and mutagenesis in λ.

Witkin presented her talk at the Bethesda meeting in October 1972. "I did not have the feeling that many in the audience were particularly convinced—the argument based on the *pol* mutant is rather subtle." In the proceedings of the meeting published in 1973, Witkin and her colleague, Donna George (not to be confused with Jacqueline George, who was associated with Devoret and later Buttin), gave full credit to Radman for the SOS hypothesis.

> Since the induction of mutations in *E. coli* by UV is also $recA^+$-dependent and lex^+-dependent, it has been suggested that bacterial UV mutability, like that of phage, may depend upon an inducible system sharing the induction requirements of prophage induction. Radman (personal communication) has developed a detailed hypothesis based on this assumption. Since UV survival and mutagenesis can only be measured in bacteria after the irradiated bacteria have divided (usually to the point of colony formation), the usual criteria for inducibility, such as the requirement for protein synthesis, cannot be applied. Therefore the hypothesis is not easily subjected to direct proof. However, we decided to consider the idea that UV-induced mutations in bacteria are caused by an inducible error-prone system (or by a system which includes one or more inducible components), which we shall call "SOS repair" (an adaptation of Radman's term "SOS replication").

Witkin began to cogitate about a more direct proof of the SOS hypothesis and recognized the utility of the temperature-sensitive *tif* mutant for inducing mutagenesis, thereby obviating the need to expose cells to levels of UV radiation that were normally required for mutation induction.

> It wasn't easy to do this. You had to expose the cells to some UV radiation; otherwise, there wouldn't be any DNA damage to serve as targets for the proposed error-prone DNA synthesis. It was quite tricky to use just enough of a tickle of UV to put in a few lesions as targets, without inducing SOS activity, and then use the high temperature as the inducing signal.

While still engaged in this work, she was invited to present a talk at the XIII International Congress of Genetics, convened in Berkeley, California in the summer of 1973. At this meeting, she presented the SOS model in a more expanded form. "I searched the literature and compiled a table of presumptive SOS functions (the first formal compilation of an ever-expanding list) known to require *recA*, *lexA*, and protein synthesis after UV irradiation, adding to those that Radman and I had already proposed. I also presented a first model of the regulation of the SOS response." The meeting was attended by Radman and Jacqueline George and the late Harrison (Hatch) Echols, whom Witkin met for the first time. "Hatch joined these discussions and became deeply interested in the SOS hypothesis from then on," she recalled. When the proceedings of the Genetics Congress were published in 1975, once again Witkin acknowledged her young colleague. "The term 'SOS repair' was introduced by Radman to convey the idea that DNA damage signals induction of a lifesaving system required to repair or cope with the damage."

Radman, still very much a junior investigator, was now languishing in the shadow of an established and well-respected scientist, his colleague (and now friend), Evelyn Witkin. By this time, he was back in Brussels as a junior faculty member and had no obvious means for access to the main pipeline of scientific visibility. His postdoctoral tenure in Meselson's laboratory had not yielded a remarkable body of work that might have provided him with greater recognition. Indeed, his attempts to isolate enzymes involved in mismatch repair and in recombination, the primary focus of his postdoctoral studies in Meselson's laboratory, had (like my own in David Goldthwait's laboratory several years earlier), led him to stumble onto a DNA repair enzyme, the significance of which was unrealized by him. Finally, with his typical reluctance to involve himself in the drudgery of writing a polished manuscript (well recognized by those who know Radman), he had not yet submitted his SOS hypothesis for formal publication, despite constant urging from Witkin to do so. From her own perspective, Witkin recounted that by late 1973:

> I was being credited for much of the SOS story, because Miro had not published his crucial ideas, except as a coauthor of the 1971 Defais et al. paper, which makes the SOS replication suggestion very briefly. I had cited Miro's unpublished memorandum in every paper I wrote, but I worried that he would not be recognized for his major part in the developing story. As a result, when I was asked to present the SOS hypothesis and the evidence for inducibility of bacterial UV mutagenesis, which by now included my direct proof using the *tif* mutant, at a conference in Rochester, I asked the organizers to invite Miro instead. I knew that the proceedings would be published and that Miro would be forced to write a paper, something I had been urging him to do. They balked at first—after all, no one had ever heard of him, and I was a well-known person in the field. I assured them that Miro would be an exciting speaker and that he should have this chance to present his ideas. The result was as I had hoped. Miro's 1974 paper is often cited as the basic SOS paper.

This paper, entitled "Phenomenology of an Inducible Mutagenic DNA Repair Pathway in *Escherichia coli*: SOS Repair Hypothesis," was indeed published in the proceedings of the Rochester meeting on *Molecular and Environmental Aspects of Mutagenesis*. In it, Radman systematically documented the (by then) weighty evidence for the existence of a number of DNA damage-inducible phenotypes under control of the *recA* and *lexA* genes. He wrote:

> I would like to propose a general working hypothesis to account for phenomena and experiments briefly reviewed in the preceding paragraphs. Chronologically, this hypothesis preceded most of the reviewed experiments for which it has already served as the working hypothesis.
>
> The principal idea is that *E. coli* possesses a DNA repair system which is repressed under normal physiological conditions but which can be induced by a variety of DNA lesions. Because of its "response" to DNA-damaging treatments we call this hypothetical repair "SOS repair." The "danger" signal which induces SOS repair is probably a temporary blockage of the normal DNA replication and possibly just the presence of DNA lesions in the cell. During the action of SOS repair mutation frequency is increased. The simplest assumption is that the SOS repair mechanism is error-prone; on the other hand, mutagenesis may be just a secondary consequence of physiological conditions under which SOS repair operates. In order

for SOS repair to function it should require some specific genetic elements, the inducing signal and *de novo* protein synthesis.

A few explanatory comments about terminology and phraseology may be necessary for the less initiated reader. The term SOS repair is now deeply ingrained in the literature. With respect to lesions such as the photoproducts that result from exposure of cells to UV radiation, the SOS response increases the amount of excision repair enzymes but does not effect a new repair pathway for such lesions. However, the SOS response does result in the repair of double-strand breaks and of blocked replication forks, and in this sense the use of the term "repair" is appropriate. The term "repair" was introduced by Witkin to connote enhanced survival of bacteria that enjoy induction of the SOS regulon. This enhanced survival is primarily the result of the ability of cells to tolerate sites of base damage by the inducible error-prone DNA synthesis mode. So, the SOS phenomenon is really a DNA damage *tolerance* rather than a DNA *repair* phenomenon. It is a cellular response mechanism whereby lesions in DNA that would otherwise block semiconservative DNA synthesis are bypassed and hence tolerated as persistent damage. The term "SOS," originally coined by Radman, is variously interpreted. Some think that he meant to imply a last-ditch attempt by the cell to survive when all other DNA repair modes had failed. What he really intended was to suggest that DNA damage was acting as a danger signal to the cell that its survival was threatened.

Witkin's experiments with the *tif* alleles of *recA* went a long way to formally proving the SOS hypothesis. Both she and the group of Jacqueline George showed that one of the thermoinducible activities resulted in elevated *spontaneous* mutability in unirradiated cells, as well as in the predicted enhancement of UV-induced mutagenesis. "SOS induction, which had been expected only to promote error-prone repair at sites of DNA damage, evidently reduced the fidelity of DNA replication on intact templates," Witkin wrote in 1982. As I indicated in the introduction to this chapter, the precise mechanism of this reduced fidelity remains a challenge. Since then, a host of fundamental experiments provided not only proof of but also mechanistic insights into the workings of the SOS phenomenon. Witkin told me that "1975 was a crucial year for the SOS story." A conference on the SOS phenomenon held in Gainesville, Florida in that year "sparked a decade of incredible activity and progress in working out the regulation and the scope of the SOS regulon." In summarizing progress in the field from 1975 to 1982, Witkin wrote the following about the Gainesville meeting.

> . . . [The] Gainesville workshop was conceived only a couple of months before it took place, and was sketchily planned over the telephone by Raymond Devoret, Ernest Pollard and myself. . . . Consequently, on April 4 [1975] about 35 participants gathered in Gainesville, to discuss the emerging evidence for a coordinately regulated group of inducible functions expressed in *E. coli* in response to DNA damage, known by then to some of us as the SOS response.
>
> On the first day of the meeting, some of the participants may well have eyed each other warily, wondering what they had in common and why they had been brought together. They represented diverse specialties, not obviously related, such as the study of prophage induction, of cell division or of mutagenesis. After three days of intense and highly synergistic interaction, almost everyone was convinced that these and other activities were, indeed, related through their common regula-

tion by the products of the two genes, *recA* and *lexA*, and by the profound changes in their expression that was triggered by DNA damage or by interruption of its replication.

Devoret, who at that time was in the United States enjoying a sabbatical year in Martin (Marty) Gellert's laboratory at the National Institutes of Health in Bethesda, vividly recalls that exciting meeting and its extraordinary synthesis of ideas and facts.

> Evelyn called me one day in the spring of 1975 and told me that she had received a call from Ernie Pollard, who was on sabbatical in Gainesville at the time. He wanted to have a meeting on lysogenic induction. She suggested that I co-organize this meeting with them and so we did and we made a list of the key people who should be invited to review the field of λ induction and of inducible mutagenesis. We invited John Clark, Jeffrey Roberts, and others, and I remember traveling to Gainsville with David Mount.

Both Roberts (from Cornell University) and Mount (from the University of Arizona) made seminal contributions to the experimental validation of the SOS hypothesis and its further elaboration. These and other important later events in the SOS story might profit the attention of a future historian.

In her 1982 *Biochimie (Paris)* paper entitled "From Gainesville to Toulouse: The Evolution of a Model," Witkin went on to elaborate that just prior to the Gainesville meeting, Jeffrey and Christine Roberts discovered that induction of phage λ was indeed accompanied by the proteolytic degradation of the λ repressor. She also recounted that at the workshop Lorraine Gudas proposed a model by which RecA and Lex A protein might regulate the SOS response. Witkin said:

> Although incorrect in some respects, the Gudas model assigned roles to the RecA and LexA proteins that have since been incorporated into the currently widely accepted and amply verified model. There had been much work by Inoue and Pardee and Lorraine Gudas, who was a graduate student at that time at Princeton with Pardee. When my *tif* paper came out showing thermal induction of the mutagenic function in *E. coli* she came to see me and asked for my *tif* strains because she had connected what she called protein X with this. And of course protein X was in fact RecA protein and they didn't know that—nobody knew that. But she was invited to the Gainesville meeting and she proposed the model that LexA protein was a repressor and that RecA protein was a proteolytically active enzyme that degraded LexA. So it was very important that she came to the meeting.

Witkin later told me that "the Gainesville meeting was extraordinary in that the people who were invited to attend held pieces of the story and had no idea that other pieces were there as well. It was a very exciting meeting from the standpoint of putting together the various pieces."

What is the evolutionary significance of the SOS phenomenon? While individual cells in a population do indeed benefit in terms of their survival, the most compelling argument about its significance that I have heard is that expressed by Miro Radman and Hatch Echols, who both suggested that the massive burst of mutations in a population of cells that undergoes SOS regulation provides the essential genetic framework in which evolutionary selection can operate. Hence, the SOS phenomenon may be thought of as a strategy for bacterial populations to adapt genetically to new and rapidly changing environmental conditions.

CHAPTER 7
Epilogue: Molecule of the Year 1994

A THOROUGH ANALYSIS of the manifold subtle and not so subtle ways in which the pursuit of scientific knowledge has become politicized in recent years is appropriately left for a different book. There is little doubt in my own mind that as we enter the 21st century, scientific journals such as *Nature*, *Science*, and *Cell*, the acknowledged "big three" in the biomedical publishing world, hold sway over a vast audience of scientific consumers, many of whom have come to equate publishing in these journals as an important if not an indispensable yardstick of scientific success. Regardless of the particular nuances that have led some members of the scientific community to accede to the attitudes and opinions of a handful of journal editors, it is now (regrettably) an undeniable fact that the imprimatur of these journals instantly defines a particular field as "hot." Such was the "fortune" of the DNA repair field when *Science* magazine identified the collective cellular DNA repair machinery as its "Molecule of the Year" in 1994.

This accolade primarily recognized two important bodies of recent scientific investigation: the establishment of a relationship between mismatch repair of DNA and colon cancer in humans, and the demonstration that some DNA repair proteins are also required for transcription by RNA polymerase II in eukaryotes. But these revelations are simply the immediate culmination of an extended period of about 15 years during which the field of DNA repair and mutagenesis has been unusually productive and informative. Horace Judson's book *The Eighth Day of Creation: Makers of the Revolution in Biology* represents an authoritative historical discourse on the period immediately following the discovery of DNA, a period accentuated chiefly by the elucidation of its function as a genetic code. The decades of the nineteen-fifties and nineteen-sixties undeniably constitute one of the golden ages of biology during which a major scientific revolution transpired. However, the history of yet another revolution in biology remains to be fully documented. The deciphering of the genetic code and the essential principles of gene regulation were followed in an astonishingly brief time by a second information explosion in biology attendant on the birth of the recombinant DNA era. A feature that distinguishes this era from the revolution that immediately preceded it is that whereas the explication of the structure of DNA heralded a period of scientific endeavor that specifically focused the intellectual spotlight on how genes function, the emergence of recombinant DNA was primarily a technological gain that provided a shot in the arm to every subdiscipline of biology endowed with investigators able and willing to exploit this technology.

The DNA repair field was most ripe for such exploitation. As earlier chapters have attempted to document, the trail emblazoned by the discovery that nature had endowed cells with multiple, diverse mechanisms for coping with natural and environmental insult to the genome began to fizzle for want of the ability to immediately define these mechanisms in detailed biochemical and molecular terms. This failure was not for any lack of effort. For reasons that likely have interesting biological implications, most cells are endowed with very small amounts of the many repair enzymes whose collective existence was so graciously recognized by *Science*. My own reflections have led me to reason that the evolutionary strategy of inventing enzymes that can compromise the integrity of the genome is something of a double-edged sword. For, if the specificity and amount of these enzymes is not rigorously controlled, they may constitute a serious threat to the viability of the DNA molecule. Regardless, countless postdoctoral fellows and graduate students were led to the sober realization that the purification of many DNA repair enzymes in amounts adequate for detailed characterization was frequently an exercise in futility until the enormous power of gene cloning and overexpression burst on to the general investigative scene. Overexpression of many recombinant proteins proved to be particularly facile in *E. coli*, rendering their purification to physical homogeneity almost trivial, and by the mid nineteen-eighties, both *E. coli* photoreactivating enzyme and the UvrA, UvrB, and UvrC proteins required for nucleotide excision repair were extensively purified and characterized. The availability of large amounts of the former protein eventually facilitated its crystallization and, as mentioned briefly at the conclusion of Chapter 2, led to a proposed structure and mechanism of action of photoreactivating enzyme. This proposed mechanism falls into the general category of so-called base flipping, by which target bases in the DNA duplex are flipped out of the helix for catalytic interactions with cognate enzymes. This mechanism of base modification appears to be common to multiple DNA repair enzymes.

The availability of the purified Uvr proteins of *E. coli* led to several profound general insights about the molecular mechanism of nucleotide excision repair. Most unexpectedly, interactions between the UvrA, UvrB, and UvrC proteins were found to lead to incision of the affected strand on both sides of sites of base damage, thereby defining an oligonucleotide fragment of a rather precise size for subsequent excision. Bimodal incision of DNA during nucleotide excision repair and the requirement for multiple interacting proteins for the recognition of damaged bases and for damage-specific incision have turned out to be general paradigms that are conserved from *E. coli* to humans. The precise number of such proteins increased profoundly during the evolution of eukaryotes. The enormous accessibility of the yeast *Saccharomyces cerevisiae* for gene cloning, coupled with the powerful selective phenotypes afforded by yeast mutants defective in nucleotide excision repair (e.g., increased sensitivity to UV radiation) led to the definition and isolation of multiple yeast genes that are required for this process. Functional cloning strategies also proved to be extraordinarily successful with respect to the isolation of human genes required for nucleotide excision repair, and led to the characterization of a series of genes clearly homologous to those isolated from yeast. A number of these genes have been definitively implicated in the human hereditary cancer-prone disease, xeroderma pigmentosum, thus providing the first unequivocal linkage between defective DNA repair and cancer predisposition. Additionally, the overexpression of genes required for nucleotide excision repair, base excision repair, and mismatch repair has facilitated the reconstitution of all these DNA repair modes in vitro in both prokaryotes and eukaryotes.

As early as the mid nineteen-eighties, the power of yeast genetics led to the provocative finding that one of the many genes required for nucleotide excision repair in yeast was also essential for the viability of haploid yeast cells. It required another decade before it was discovered that this essential role derives from a second function of the protein encoded by this gene, namely its participation in the initiation of transcription by RNA polymerase II. The simultaneous discovery that a different nucleotide excision repair protein in human cells was also required for transcription initiation led to a flood of studies which culminated in the identification of even more proteins that are indispensable for nucleotide excision repair in eukaryotes and that are also essential for transcription initiated by RNA polymerase II. Why are so many proteins required for the early steps of nucleotide excision repair in eukaryotic cells, and what is the biological significance of the dual role of some of these proteins in nucleotide excision repair and transcription? Definitive answers to these questions constitute some of the many challenges for the future.

Other relationships between nucleotide excision repair and transcription have emerged. Prompted by an interest in exploring the possible heterogeneity of DNA repair in different regions of the human genome, Philip Hanawalt and his colleagues observed that in *E. coli* as well as in human cells the transcribed strand of transcriptionally active genes undergoes nucleotide excision repair at a significantly faster rate than the nontranscribed strand. In *E. coli*, this kinetic preference appears to reflect the rapid targeting of the nucleotide excision repair apparatus to sites of stalled transcription via a specific coupling factor(s). It is likely that a functionally analogous coupling factor(s) exists in eukaryotes. To date, the biological significance of so-called strand-specific preferential repair is unclear and this particular connection between nucleotide excision repair and transcription appears to be unrelated to the dual function of the nucleotide excision repair/transcription proteins discussed above.

Regardless of its ultimate utility to the cell, the fact that some nucleotide excision repair proteins are also required for normal transcription initiation has unlocked a new area of investigative inquiry which is focused on the potential pathogenetic consequences of defective transcription initiation. Interest in this arena has been heightened by the direct implication of several nucleotide excision repair/transcription genes in the pathology of a number of hereditary human syndromes. These syndromes, of which the diseases trichothiodystrophy and a form of Cockayne syndrome associated with xeroderma pigmentosum are prime examples, are characterized by complex clinical phenotypes that include abnormal postnatal growth and development as well as neurological deficiencies. Cockayne syndrome can also occur without the attendant clinical features of xeroderma pigmentosum. Two genes implicated in this form of the disease have been isolated. The function of the polypeptides they encode is unknown. However, if the transcription hypothesis of Cockayne syndrome associated with xeroderma pigmentosum turns out to be validated, the argument is compelling that defective transcription is also the essential factor underlying the pathogenesis of the "pure" form of Cockayne syndrome. Herein lies yet another possible relationship between DNA repair and transcription because the cellular phenotype of the uncomplicated form of Cockayne syndrome includes a sensitivity to UV radiation as well as defective strand-specific nucleotide excision repair.

As indicated above, another clear relationship has emerged between the failure to repair damaged DNA and a predisposition to cancer. Individuals who are genetically defective in any of several genes that encode proteins required for the repair of mispaired bases are highly prone to cancer of the colon, and sometimes of other organs. The extraordinary frequency of colon cancer in the general population thus identifies DNA repair genes as important oncogenes. Abnormal cellular responses to DNA damage have been implicated in several other hereditary diseases, including Fanconi anemia, ataxia telangiectasia, Li-Fraumeni syndrome, the so-called 46BR syndrome, and Bloom syndrome, though it is not at all certain that defective DNA repair underlies these abnormal responses. Nonetheless, genes involved in all of these diseases have recently been cloned. Indeed, at the time of writing, genes for some of the Fanconi anemia genetic complementation groups and for two of the xeroderma pigmentosum genetic complementation groups are the only ones remaining to be isolated among this general group of human diseases. Surprisingly, hereditary human diseases or genetic cancer predispositions related to defects in proteins required for base excision repair have not yet been identified. The systematic examination of human tumors carrying homozygous mutations in genes that encode such proteins might prove a fruitful way of uncovering such syndromes. In the same vein, continued surveillance for patients that have unexpectedly severe reactions to radiotherapy may be a productive strategy for uncovering diseases other than ataxia telangiectasia with a cellular phenotype of abnormal sensitivity to agents that induce strand breaks in DNA.

The gene for Bloom's syndrome (*BLM*) and that for ataxia telangiectasia (*ATM*) together with the various nucleotide excision repair, mismatch repair, and Cockayne syndrome genes just mentioned, provide an impressive armada of human disease genes that can be readily translated to their mouse homologs for the generation of animal models of these hereditary diseases. "Knock-out" mouse strains are now available that are defective in the xeroderma pigmentosum group A (*XPA*) and group C (*XPC*) genes and in a nucleotide excision repair gene called *ERCC1*. Strains have also been constructed that are defective in the Cockayne syndrome group A (*CSA*) and group B (*CSB*) genes and in several different mismatch repair genes. Aside from their enormous potential for deciphering the molecular pathogenesis of these particular disease states in humans, these mutant mouse strains offer unparalleled opportunities for dissecting the genetics of cancer in mammals by mating them to one another and to strains carrying mutations in other known oncogenes and tumor suppressor genes which are not directly related to DNA repair. Surprises from such genetic crosses are already beginning to emerge. For example, when homozygous mutant XP-A or XP-C mice are mated to mice carrying homozygous mutations in the *CSB* gene, there is a profound growth phenotype. The potential for generating mouse strains that are defective in all three primary excision repair modes, nucleotide excision repair, base excision repair, and mismatch repair, offers opportunities to evaluate the total contribution of DNA repair processes to spontaneous and DNA-damage-induced carcinogenesis and to growth and development.

The elucidation of interactions between the cellular DNA repair machinery and biochemical events that regulate the cell cycle is another recent exciting area of research. Examination of the eukaryotic cell cycle led to the early insight that eukaryotic cells have the capacity for regulating and checking the accurate completion of each cell cycle at specific checkpoints, before proceeding to the next. This checking procedure includes the capacity for not only ensuring that each critical phase of the cycle has been completed, but also that the integrity of the genome has not been compromised, including through insult by environmental DNA-damaging agents. There are indications that these checkpoints operate in the G1, S, and G2 phases of the cell cycle and result in regulated arrest of the cycle, thereby providing kinetic windows that allow for increased opportunities for DNA repair mechanisms to operate. There are also indications that the intervention of DNA repair at appropriate opportunities is not simply a passive affair, but is somehow activated as part of the complex checkpoint control mechanism. Considerable interest derives from the fact that the tumor suppressor gene p53 is intimately tied into this control system, as is the *ATM* gene responsible for the human hereditary disease ataxia telangiectasia. The study of checkpoint control in the cell cycle is currently extremely active, and holds promise not only of providing new insights into mechanisms of neoplastic transformation, but also of novel chemotherapeutic anticancer drug design.

A final area of cellular responses to DNA damage that has enjoyed spectacular recent progress is that of double-strand break repair in mammalian cells. It has long been known that in such cells DNA double-strand breaks are mediated by some type of end-to-end fusion process rather than by homologous recombination, as is the case in yeast. Details of this repair mechanism are still obscure, but the association between strand break repair and a previously characterized enzyme called DNA-dependent protein kinase (DNA-PK) represents a major breakthrough. DNA-PK is comprised of the so-called Ku autoantigens, which bind to the ends of breaks as a heterodimer, and a catalytic subunit that acquires a kinase activity upon binding to DNA-bound Ku proteins. Indeed, mutants have been found for all subunits in human and rodent cells that are sensitive to ionizing radiation and defective in double-strand break repair and V(D)J recombination. The latter is a process that promotes fusion between constant and variable regions of genes that encode antibodies during immune cell differentiation and is initiated by a site-specific DNA double-strand break.

In conclusion, I think that it is safe to state that the field of DNA repair and mutagenesis, more properly the field of cellular responses to genomic injury and insult, has matured enormously since 1949, when the discovery of photoreactivation was formally announced. Even in the (likely) possibility that there are in fact no fundamentally new DNA repair mechanisms to be uncovered in nature, the future of the discipline is secured in mainstream molecular and cellular biology by dint of its firmly established central role in neoplastic transformation, cell-cycle regulation, transcription, and DNA replication. It is my hope and my expectation that future historians of this topic will have a difficult time documenting the development of this field in the relative isolation from the rest of mainstream molecular biology that I have attempted.

Notes

Complete titles of books and journals referred to frequently are given on first mention in each chapter and abbreviated thereafter.

CHAPTER 1

1. Judson—Horace Freeland Judson, *The Eighth Day of Creation: Makers of the Revolution in Biology* (New York: Simon and Schuster, 1979)[†], p. 48.
1. Muller's paper—Hermann J. Muller, "Induced Mutations in Drosophila," *Cold Spring Harbor Symposia on Quantitative Biology* 9 (1941): 151–167.
2. discovery of UV radiation—John Jagger, *Introduction to Research in Ultraviolet Photobiology* (Engelwood Cliffs, New Jersey: Prentice Hall, 1967), p. 1.
2. observation of Downes and Blunt—L. R. Koller, "Bactericidal Effects of Ultraviolet Radiation Produced by Low Pressure Mercury Vapor Lamps," *Journal of Applied Physics* 10 (1939): 624–630.
2. early experimentation with UV radiation—Jagger, *Introduction to Research in Ultraviolet Photobiology*, p. vii.
3. introductory remarks by Harris at the first Cold Spring Harbor Symposium—Reginald G. Harris, "Introduction," *C. S. H. Symp. Q. Biol.* 1 (1933): v–vi.
3. historical reflections of Symonds—Neville Symonds, "Schrödinger and Delbrück: Their Status in Biology," *Trends in Biochemical Sciences* 13 (1988): 232–234.
4. Kimball's review—Richard Kimball, "The Development of Ideas about the Effect of DNA Repair on the Induction of Gene Mutations and Chromosomal Aberrations by Radiation and Chemicals," *Mutation Research* 186 (1987): 1–34.
4. demonstration that the genetic material is DNA—Oswald T. Avery, Colin M. MacLeod, and Maclyn McCarty, "Studies on the Chemical Nature of the Substance Inducing Transformation of Pneumococcal Types. Induction of Transformation by a Desoxyribonucleic Acid Fraction Isolated from Pneumococcus Type II T," *Journal of Experimental Medicine* 79 (1944): 137–158.
6. Hollaender's paper—Alexander Hollaender, "History of Radiation Biology from a Personal Point of View," in *Mechanisms of DNA Damage and Repair*, ed. M. G. Simic, L. Grossman, and A. C. Upton (New York: Plenum, 1986), pp. 9–17.

[†]All page numbers cited refer to this edition; however, the interested reader will note that the expanded edition is now available (Cold Spring Harbor, New York: Cold Spring Harbor Laboratory Press, 1996).

7 Setlow—Conversation with Dick Setlow, June 6, 1995.
7 von Borstel—Conversation with Jack von Borstel, May 25, 1995.
8 observation that "the wavelengths of light most effective for killing bacteria were those that were most efficiently absorbed by nucleic acids"—Frederick L. Gates, "On Nuclear Derivatives and the Lethal Action of Ultraviolet Light," *Science* 68 (1928): 479–480.
8 Jagger's comments on Gates's observation—J. W. Longworth, J. Jagger, and W. Shropshire, Jr., eds., *Photobiology 1984* (New York: Praeger, 1985), pp. 123–133.
8 similarity between UV action spectrum and absorption spectrum of nucleic acids—Alexander Hollaender and C. W. Emmons, "Wavelength Dependence of Mutation Production in the Ultraviolet with Special Emphasis on Fungi," *C. S. H. Symp. Q. Biol.* 9 (1941): 179–186.
9 Muller's comments on Hollaender's results—Hermann J. Muller, "Résumé and Perspectives of the Symposium on Genes and Chromosomes," *C. S. H. Symp. Q. Biol.* 9 (1941): 290–308.
9 discussion with Max Delbrück—Judson, *The Eighth Day of Creation*, p. 59.
9 McCarty's recollections—Maclyn McCarty, in *The Transforming Principle: Discovering That Genes Are Made of DNA* (New York: W.W. Norton, 1985), p. 215.
10 use of microorganisms for genetic studies—Judson, *The Eighth Day of Creation*, p. 55.
10 reference to Alfred Mirsky—Judson, *The Eighth Day of Creation*, p. 40.
10 Haynes—Conversation with Bob Haynes, March 27, 1995.
11 discussion of Delbrück's work—Judson, *The Eighth Day of Creation*; Ernst Peter Fischer and Carol Lipson, *Thinking About Science: Max Delbrück and the Origins of Molecular Biology* (New York: W.W. Norton, 1988), p. 79.
12 excerpt from Bohr's lecture and Delbrück's interest in complementarity—Fischer and Lipson, *Thinking About Science*, p. 79.
12 Zimmer's recollections—Fischer and Lipson, *Thinking About Science*, p. 88.
12 quantum model of gene mutation—Fischer and Lipson, *Thinking About Science*, p. 89.
12 Witkin's reflections about mutagenesis—Evelyn M. Witkin, "Ultraviolet Mutagenesis and the SOS Response in *Escherichia coli*: A Personal Perspective," *Environmental and Molecular Mutagenesis* 14 (1989): Suppl. 16, 30–34.
13 Schrödinger—Erwin Schrödinger, *What Is Life? The Physical Aspect of the Living Cell* (Cambridge: At the University Press, 1944).
13 Welch's analysis—G. R. Welch, "Schrödinger's *What Is Life*: A 50-year Reflection," *Trends Biochem. Sci.* 20 (1995): 45–48.
13 influence of Schrödinger's book on Watson—Judson, *The Eighth Day of Creation*, p. 47; J. D. Watson, "Succeeding in Science: Some Rules of Thumb," *Science* 261 (1993): 1812–1813.
14 Delbrück's letter to the Rockefeller Foundation—Fischer and Lipson, *Thinking About Science*, p. 96.
14 Timoféeff-Ressovsky, biographical details—Fischer and Lipson, *Thinking About Science*, p. 98.
14 Delbrück's early years in the U.S. and Ellis's comments—Judson, *The Eighth Day of Creation*, p. 51–52.
15 Strauss—Conversation with Bernie Strauss, June 29, 1995.
16 Auerbach's discovery of the mutagenic effects of mustard gas—Charlotte Auerbach, "Problems in Chemical Mutagenesis," *C. S. H. Symp. Q. Biol.* 16 (1951): 199–214.
16 tribute to Auerbach Lotte—B. J. Kilby, "In Memorium—Charlotte Auerbach, FRS (1899–1994)," *Mutat. Res.* 327 (1995): 1–4.

16 Haynes—Conversation with Bob Haynes, March 27, 1995.
17 Mazia—Daniel Mazia, "Physiology of the Cell Nucleus," in E. G. Barron, ed., *Modern Trends in Physiology and Biochemistry* (New York: Academic Press, 1952), pp. 77–123.
17 Stahl—Letter from Franklin Stahl, February 21, 1995.
17 Nadson—G. A. Nadson, "On Action of Radium on Yeast Fungi in Connection with the General Problem of Radium Effect on Living Matter," *Vestnik Rentgenologii i Radiologii* 1 (1920): 45–137.
17 Roberts and Aldous—Richard B. Roberts and Elaine Aldous, "Recovery from Ultraviolet Irradiation in *Escherichia coli*," *Journal of Bacteriology* 57 (1949): 363–375.
18 Bovie's work—Excerpts from Paul Swenson's unpublished manuscript courtesy of Dr. Claud S. (Stan) Rupert.
18 Hollaender and Curtis, radiation and growth lag—J. Curtis and A. Hollaender, "Effect of Sublethal Doses of Monochromatic Ultraviolet Radiation on Bacteria in Liquid Suspensions," *Proceedings of the Society for Experimental Medicine* 33 (1935): 61–62.
18 Hollaender and Emmons, recovery of irradiated spores—C. W. Emmons and A. Hollaender, "The Action of Ultraviolet Radiation on Dermatophytes. II. Mutations Induced in Cultures of Dermatophytes by Exposure of Spores to Monochromatic Ultraviolet Radiation," *American Journal of Botany* 26 (1939): 467–475.
18 Hollaender and Emmons, variations in the yield of UV radiation-induced mutations—Cited previously, note to p. 8.
18 foreword to the Squaw Valley symposium proceedings—A. Hollaender, in *Molecular Mechanisms for Repair of DNA* (New York: Plenum Press, 1975), part A, pp. vii–viii.
19 Latarjet—Raymond Latarjet, "Action du froid sur la réparation des radio-lésions chez une levure et chez une bactérie," *Comptes rendus Hebdomadaires des séances de l'Académie des Sciences* 217 (1943): 186–188.
19 Roberts and Aldous's report—Cited previously, note to p. 17.
20 excerpts from Luria's autobiography regarding his medical training—Salvador E. Luria, *A Slot Machine, A Broken Test Tube: An Autobiography* (New York: Harper and Row, 1984), p. 14 and p. 17.
21 "inner circle"—Judson, *The Eighth Day of Creation*, p. 191.
21 "Holy Grail of biophysics"—Luria, *A Slot Machine, A Broken Test Tube*, p. 20.
21 "trolley-car accident"—Luria, *A Slot Machine, A Broken Test Tube*, p. 20.
22 analysis of the properties and growth of bacteriophages with UV radiation—S. E. Luria, "Reactivation of Ultraviolet-irradiated Bacteriophage by Multiple Infection," *Journal of Cellular and Comparative Physiology* 39 (1952): Suppl. 1, 119–123.
22 Luria on DNA repair phenomena—Luria, *A Slot Machine, A Broken Test Tube*, pp. 96–97.
23 reference to multiplicity reactivation—Judson, *The Eighth Day of Creation*, p. 66.
23 comments regarding Watson's Ph.D. thesis—Judson, *The Eighth Day of Creation*, p. 68.
24 focus on phages T 1–7—Fischer and Lipson, *Thinking About Science*, p. 154.
24 phage T4 was more resistant to inactivation by UV radiation than T2 or T6—S. E. Luria, "Reactivation of Irradiated Bacteriophage by Transfer of Self-Reproducing Units," *Proceedings of the National Academy of Sciences* 33 (1947): 253–264.
25 literary note—Nathaniel Wanley, 1678. *The Wonders of the Little World*.

CHAPTER 2

28 Kelner—Conversation with Adelyn Kelner, June 30, 1995.
28 Rupert—Conversations with Stan Rupert, March 15, April 18, and September 12, 1995.
29 Watson's recollection of Demerec—Horace Freeland Judson, *The Eighth Day of Creation: Makers of the Revolution in Biology* (New York: Simon and Schuster, 1979), p. 66.
30 Kelner's account of his work—Albert Kelner's letter to Stan Rupert written in late 1961.
34 Letters to Luria from Kelner—courtesy of Adelyn Kelner.
40 Dulbecco's paper describing photoreactivation in phage—Renato Dulbecco, "Reactivation of Ultraviolet-inactivated Bacteriophage by Visible Light," *Nature* 163 (1949): 949–950.
40 Luria's autobiography—Salvador E. Luria, *A Slot Machine, A Broken Test Tube: An Autobiography* (New York: Harper and Row, 1984).
42 Dulbecco—Conversation with Renato Dulbecco, August 15, 1995.
42 Dulbecco's suggestion that the human genome be sequenced—R. Dulbecco, "A Turning Point in Cancer Research: Sequencing the Human Genome," *Science* 231 (1986): 1055–1056.
45 Kelner's paper presented at the Research Conference for Biology and Medicine of the Atomic Energy Commission, 1949—A. Kelner, "Experiments on Photoreactivation with Bacteria and Other Microorganisms," *Journal of Cellular and Comparative Physiology* 39 (1952): Suppl. 1, 115–117.
45 Delbrück's biography—Ernst Peter Fischer and Carol Lipson, *Thinking About Science: Max Delbrück and the Origins of Molecular Biology* (New York: W.W. Norton, 1988).
46 Delbrück's letter to Luria—Fischer and Lipson, *Thinking About Science*, p. 184.
46 Novick and Szilard's observations—A. Novick and L. Szilard, "Experiments on Light-reactivation of Ultra-violet Inactivated Bacteria," *Proceedings of the National Academy of Sciences* 25 (1949): 591–600.
47 UV-induced inhibition of DNA synthesis—A. J. Kelner, "Growth, Respiration, and Nucleic Acid Synthesis in Ultraviolet-irradiated and in Photoreactivated *Escherichia coli*," *Journal of Bacteriology* 65 (1953): 252–262.
47 Cairns—Correspondence with John Cairns, October 1995.
48 Probable earliest documented use of the term "repair" with reference to physiological responses to genomic injury—S. E. Luria, "Reactivation of Ultraviolet-irradiated Bacteriophage by Multiple Infection," *J. Cell. Comp. Physiol.* 39: (1952) Suppl. 1, 119–123.
48 absorption of light must be effected by specific light-accepting moieties in cells—A. Kelner, "Action Spectra for Photoreactivation of Ultraviolet-irradiated *Escherichia coli* and *Streptomyces griseus*," *Journal of General Physiology* 34 (1949): 835–852.
48 Dulbecco's further studies on photoreactivation—R. Dulbecco, "Experiments on Photoreactivation of Bacteriophages Inactivated with Ultraviolet Radiation," *J. Bacteriol.* 59 (1950): 329–347.
48 Watson and Crick—J. D. Watson and F. H. C. Crick, "Genetical Implications of the Structure of Deoxyribonucleic Acid," *Nature* 171 (30 May 1953): 964–967.
48 Judson's comments regarding Crick and Watson—Judson, *The Eighth Day of Creation*, p. 126.
49 Crick's comments—Francis Crick, *What Mad Pursuit: A Personal View of Scientific Discovery* (New York: Basic Books, 1988), p. 111.
50 Goodgal—Conversation with Sol Goodgal, July 13, 1995.

50 Herriot's letter to Hershey—Judson, *The Eighth Day of Creation*, p. 57.
53 paper by Rupert and colleagues—C. S. Rupert, S. Goodgal, and R. M. Herriott, "Photoreactivation In Vitro of Ultraviolet Inactivated *Haemophilus influenzae* Transforming Factor," *J. Gen. Physiol.* 41 (1958): 451-471.
53 Crick's comments—F. H. C. Crick, "The Double Helix: A Personal View," *Nature* 248 (1974): 766-769.
54 Judson's remarks—Judson, *The Eighth Day of Creation*, p. 447.
54 Benzer's goal—Judson, *The Eighth Day of Creation*, p. 271.
54 first UV radiation-sensitive mutant of *E. coli*—R. F. Hill, "A Radiation-sensitive Mutant of *Escherichia coli*," *Biochimica et Biophysica Acta* 30 (1958): 636-637; and S. A. Ellison, R. R. Feiner, and R. F. Hill, "A Host Effect on Bacteriophage Survival after Ultraviolet Irradiation," *Virology* 11 (1960): 294-296.
55 Straub—Fischer and Lipson, *Thinking About Science*, p. 264.
55 Delbrück's letter to Beadle—Fischer and Lipson, *Thinking About Science*, p. 267.
56 comments regarding Delbrück's institute at the University of Cologne—Fischer and Lipson, *Thinking About Science*, pp. 268-269.
56 Delbrück's letter to the dean—Fischer and Lipson, *Thinking About Science*, p. 271.
57 Hillebrandt and Harm—W. Harm and B. Hillebrandt, "A Non-photoreactivable Mutant of *E. coli* B," *Photochemistry and Photobiology* 1 (1962): 271-272.
57 Rupert's studies with *E. coli*—C. S. Rupert, "Relation of Photoreactivation to Photoenzymatic Repair of DNA in *Escherichia coli*," *Photochem. Photobiol.* 4 (1965): 271-275; and C. S. Rupert, "Photoreactivation of Ultraviolet Damage," *Photophysiology* 2 (1964): 283-327.
57 Rupert's studies with *S. cerevisiae* and comments regarding "the [remaining] problem"—C. S. Rupert, "Photoreaction of Transforming DNA by an Enzyme from Baker's Yeast," *J. Gen. Physiol.* 43 (1960): 573-595.
57 Szent-Gyorgi's observations—S. Y. Wang, "Photochemical Reactions in Frozen Solutions," *Nature* 190 (1961): 690-694.
58 work of Beukers and his colleagues—R. Beukers, J. Ijlstra, and W. Berends, "The Effect of Ultraviolet Light on Some Components of the Nucleic Acids II. In Rapidly Frozen Solutions," *Recueil des Travaux Chimiques du Pays-bas et de la Belgique* 77 (1958): 729-732; and R. Beukers and W. Berends, "Isolation and Identification of the Irradiation Product of Thymine," *Biochim. Biophys. Acta* 41 (1960): 550-551.
58 Wang's paper—S. Y. Wang, "Photochemical Reactions in Frozen Solutions," *Nature* 190 (1961): 690-694.
58 observations by the Dutch group—R. Beukers, J. Ijlstra, and W. Berends, "The Effect of Ultraviolet Light on Some Components of the Nucleic Acids VI. The Origin of the UV Sensitivity of Deoxyribonucleic Acid," *Recl. Trav. Chim. Pays-bas Belg.* 79 (1960): 101-104.
59 the Setlows' work—R. B. Setlow, "The Action Spectrum for the Reversal of the Dimerization of Thymine Induced by Ultraviolet Light," *Biochim. Biophys. Acta* 49 (1961): 237-238; R. B. Setlow and W. L. Carrier, "Identification of Ultraviolet-induced Thymine Dimers in DNA by Absorbance Measurements," *Photochem. Photobiol.* 2 (1963): 49-57; R. B. Setlow and J. K. Setlow, "Evidence that Ultraviolet-induced Thymine Dimers in DNA Cause Biological Damage," *Proc. N. A. S.* 48 (1962): 1250-1257.
59 Dick Setlow—Conversation with Dick Setlow, June 6, 1995.
59 Rupert and Wulff—D. L. Wulff and C. S. Rupert, "Disappearance of Thymine Photodimer in Ultraviolet Irradiated DNA upon Treatment with a Photoreactivating Enzyme from Baker's Yeast" *Biochemical and Biophysical Research Communications* 7 (1962): 237-240.

CHAPTER 3

63 Haynes—Conversation with Bob Haynes, March 27, 1995.
63 Watson and Crick—J. D. Watson and F. H. C. Crick, "A Structure for Deoxyribose Nucleic Acid," *Nature* 171 (1953): 737–738.
64 article by Haynes and Hanawalt—P. C. Hanawalt and R. B. Haynes, "The Repair of DNA," *Scientific American* 216 (1967): 36–43.
65 tribute to Hill—R. H. Haynes, in *Molecular Mechanisms for Repair of DNA*, ed. P. C. Hanawalt and R. B. Setlow, *Basic Life Sciences*, vol. 5A (New York: Plenum Press, 1975), part A, front matter.
65 Harm's contribution—Correspondence from Walter Harm, March 1, 1995 and November 6, 1995.
65 nomenclature—E. C. Friedberg, *DNA Repair* (New York: W.H. Freeman, 1984).
66 Sauerbier—W. Sauerbier, "Evidence for a Nonrecombinational Mechanism of Host Cell Reactivation of Phage," *Virology* 16 (1962): 398–404.
66 Strauss—Correspondence from Bernard Strauss, September 5, 1995.
67 Meselson on Setlow and Hollaender—Conversation with Matthew Meselson, September 15, 1995.
67 Setlow—Conversation with Dick Setlow, June 6, 1995.
67 Witkin's recollections—E. M. Witkin, "Mutation Frequency Decline Revisited," *BioEssays* 16 (1994): 437–444.
69 Witkin—Conversations with Evelyn Witkin, May 22 and July 25, 1995.
69 Witkin's recollections—Cited previously, note to p. 67.
70 Witkin's paper—E. M. Witkin, "Time, Temperature, and Protein Synthesis: A Study of Ultraviolet-induced Mutation in Bacteria," *Cold Spring Harbor Symposia on Quantitative Biology* 21 (1956): 123–140.
71 Witkin's challenge to Setlow—Conversations with Evelyn Witkin, May 22 and July 25, 1995; conversation with Dick Setlow, June 6, 1995.
72 Haynes's autobiographical discourse—Robert H. Haynes, "My Road to Repair in Yeast: The Importance of Being Ignorant," in *The Early Days of Yeast Genetics*, ed. M. N. Hall and P. Linder (Cold Spring Harbor, New York: Cold Spring Harbor Laboratory Press, 1993), pp. 145–171.
73 Haynes's recollections—Cited previously, note to p. 72.
73 Korogodin's review—V. I. Korogodin, "The Study of Post-Irradiation Recovery of Yeast: The 'Premolecular Period'," *Mutation Research* 289 (1993): 17–26.
73 Sherman and Chase's paper—F. G. Sherman and H. B. Chase, "Effects of Ionizing Radiations on Enzymatic Activities of Yeast Cells. I. Relation Between Anaerobic CO_2 Production and Colony Production at Intervals After X-radiation," *Journal of Cellular and Comparative Physiology* 33 (1949): 17–23.
74 Hanawalt—Conversations with Phil Hanawalt, August 2 and September 24, 1995.
75 Cairns—Correspondence from John Cairns, October 15, 1995.
75 Harm's admission—Correspondence from Walter Harm, November 6, 1995.
75 Roberts and Aldous' seminal paper—Richard B. Roberts and Elaine Aldous, "Recovery from Ultraviolet Irradiation in *Escherichia coli*," *Journal of Bacteriology* 57 (1949): 363–375.
75 Haynes's historical review—Cited previously, note to p. 72.
76 Haynes and Patrick's liquid holding recovery findings—M. H. Patrick, R. H. Haynes, and R. B. Uretz, "The Possibility of Repair of Primary Radiation Damage in Yeast" (abstract), *Radiation Research* 16 (1962): 610.
76 Haynes on the historiana of excision repair—Conversation with Bob Haynes, March 27, 1995.
77 Haynes's paper with Inch—R. H. Haynes and W. R. Inch, "Synergistic Action of

Nitrogen Mustard and Radiation in Microorganisms," *Proceedings of the National Academy of Sciences* 50 (1963): 839–846.

77 Guild's paper—W. R. Guild, "The Radiation Sensitivity of Deoxyribonucleic Acid," *Radiat. Res.* Suppl. 3 (1963): 257–269.

77 Haynes's summation of events—Letter from Bob Haynes, May 25, 1995.

78 Judson's comments—Horace Freeland Judson, *The Eighth Day of Creation: Makers of the Revolution in Biology* (New York: Simon and Schuster, 1979), p. 321.

79 reception to Volkin and Astrachan's findings—Judson, *The Eighth Day of Creation*, p. 325.

81 Setlow group's paper—R. B. Setlow, P. A. Swenson, and W. L. Carrier, "Thymine Dimers and the Inhibition of DNA Synthesis by UV Irradiation of Cells," *Science* 142 (1963): 1464–1466.

82 Setlow and Carrier's classic paper—R. B. Setlow and W. L. Carrier, "The Disappearance of Thymine Dimers From DNA: An Error-Correcting Mechanism," *Proc. N. A. S.* 51 (1964): 226–231.

84 announcement of excision repair—Cited previously, note to p. 82.

84 Setlow's summary of symposium—R. B. Setlow, "Summary," *Radiat. Res.* Suppl. 6 (1966): 220–226.

84 Setlow and Carrier's speculation on generality of excision repair—Cited previously, note to p. 82.

89 Hanawalt's historical reflections—P. C. Hanawalt, "Concepts and Models for DNA Repair: From *E. coli* to Mammalian Cells," *Environmental and Molecular Mutagenesis* 14 (1989): Suppl. 16, 90–98.

90 Setlow's note—Letter from Dick Setlow to Phil Hanawalt, August 23, 1963, courtesy of Phil Hanawalt.

90 Hanawalt and Pettijohn's definitive exposition—D. Pettijohn and P. Hanawalt, "Evidence for Repair-Replication of Ultra-Violet Damaged DNA in Bacteria," *Journal of Molecular Biology* 9 (1964): 395–410.

91 Boyce's account of the excision repair story—Unpublished narrative from Richard P. Boyce with accompanying correspondence dated June 1, 1995.

93 Kahn's book—Carol Kahn, *Beyond the Helix: DNA and the Quest for Longevity* (New York: Times Books, 1985)

94 Waldstein's version—Conversation with Evelyn Waldstein, September 24, 1995.

96 Witkin's important emphasis—E. M. Witkin, "Mutation and the Repair of Radiation Damage in Bacteria," *Radiat. Res.* Suppl. 6 (1966): 30–51.

97 Painter's observation—R. B. Painter, "On Tritium, DNA, and Serendipity," *Radiat. Res.* 121 (1990): 117–119.

97 Setlow at the Chicago meeting—Cited previously, note to p. 84.

97 Painter and Rasmussen's paper—R. E. Rasmussen and R. B. Painter, "Evidence for Repair of Ultra-Violet Damaged Deoxyribonucleic Acid in Cultured Mammalian Cells," *Nature* 203 (1964): 1360–1362.

97 Painter's recollection—Undated correspondence from Robert Painter circa mid 1995.

98 Cleaver's recollections—Conversation with James Cleaver, May 23, 1995.

99 Cleaver's fascinating observations—James Cleaver, "Defective Repair Replication of DNA in Xeroderma Pigmentosum," *Nature* 218 (1968): 652–656.

99 Lederberg's editorial—Joshua Lederberg, "That 'Academic' Work on DNA Applied to a Human Disease," *The Washington Post*, 8 June 1968.

100 Haynes and Hanawalt on their work with nitrogen mustard—Conversation with Bob Haynes, March 27, 1995; P. C. Hanawalt, cited previously (note to p. 89).

CHAPTER 4

114 Sekiguchi—Correspondence with Mutsuo Sekiguchi, September 1, 1995.
114 Strauss—Conversation with Bernie Strauss, June 25, 1995.
117 Grossman—Correspondence with Lawrence Grossman, September 18, 1995.
120 Lindahl—Correspondence with Tomas Lindahl, August 11, 1995.
120 Lindahl's comments—T. Lindahl, "An N-Glycosidase from *Escherichia coli* That Releases Free Uracil from DNA Containing Deaminated Cytosine Residues," *Proceedings of the National Academy of Sciences* 71 (1974): 3649–3653.
121 Lindahl's review—T. Lindahl, "DNA Glycosylases, Endonucleases for Apurinic/Apyrimidinic Sites, and Base Excision Repair," *Progress in Nucleic Acid Research* 22 (1979): 135–192.
122 Lindahl's paper—Cited previously, note to p. 120.
123 Judson's suggestion to Crick—Horace Freeland Judson, *The Eighth Day of Creation: Makers of the Revolution in Biology* (New York: Simon and Schuster, 1979), p. 180.
123 Crick's response—Judson, *The Eighth Day of Creation*, p. 181.
123 Judson's comments—Judson, *The Eighth Day of Creation*, p. 93.
123 Judson's discussion of Chargaff's discovery—Judson, *The Eighth Day of Creation*, p. 96.
125 excerpt from Friedberg's textbook—E. C. Friedberg, *DNA Repair* (New York: W.H. Freeman, 1984), p. 154.
126 endonuclease in *E. coli* that hydrolyzes apurinic sites in DNA—W. G. Verly and Y. Paquette, "An Endonuclease for Depurinated DNA in *Escherichia coli* B," *Canadian Journal of Biochemistry* 50 (1972): 217–224.
126 Lindahl's paper—T. Lindahl and A. Andersson, "Rate of Chain Breakage at Apurinic Sites in Double-Stranded Deoxyribonucleic Acid," *Biochemistry* 11 (1972): 3618–3623.
130 Cairns—Correspondence with John Cairns, October 1995.
132 excerpts from the Leeuwenhoek Lecture to the Royal Society—J. Cairns, "Bacteria as Proper Subjects for Cancer Research," *Proceedings of the Royal Society of London - Series B: Biological Sciences* 208 (1980): 121–133.
133 Samson—Conversation with Leona Samson, June 5, 1995.
134 adaptive response to alkylation damage—L. Samson and J. Cairns, "A New Pathway for DNA Repair in *Escherichia coli*," *Nature* 267 (1977): 281–283.
136 Cairns's thoughts about adaptation—Letters from John Cairns to Paul Schendel, courtesy of Dr. John Cairns.
137 Letters to *Nature*—P. Robins and J. Cairns, "Quantitation of the Adaptive Response to Alkylating Agents," *Nature* 280 (1979): 74–76; P. Karran, T. Lindahl, and B. Griffin, "Adaptive Response to Alkylating Agents Involves Alteration *in situ* of O6-Methylguanine Residues in DNA," *Nature* 280 (1979): 76–77.
137 Lindahl and Karran's paper—See Karran et al. 1979 cited above.

CHAPTER 5

140 Lacks—Conversation with Sandy Lacks, November 28, 1995.
140 Judson's comments—Horace Freeland Judson, *The Eighth Day of Creation: Makers of the Revolution in Biology* (New York: Simon and Schuster, 1979), p. 93.
141 Ephrussi-Taylor's paper—H. Ephrussi-Taylor and T. C. Gray, "Genetic Studies of Recombining DNA in Pneumococcal Transformation," *Journal of General Physiology* 49 (1966): part II, 211–231.

143 Lacks's paper—S. Lacks, "Mutants of *Diplococcus pneumoniae* That Lack Deoxyribonucleases and Other Activities Possibly Pertinent to Genetic Transformation," *Journal of Bacteriology* 101 (1970): 373–383.
144 Fox—Conversation with Maurice Fox, January 12, 1996.
144 excerpt from Lacks's paper—Cited previously, note to p. 143.
144 Tiraby and Fox's paper—J.-G. Tiraby and M. S. Fox, "Marker Discrimination of Transformation and Mutation of Pneumococcus," *Proceedings of the National Academy of Sciences* 70 (1973): 3541–3545.
144 Radman—Conversation with Miroslav Radman, October 5, 1995.
145 Whitehouse's proposal—H. L. K. Whitehouse, "A Theory of Crossing-Over by Means of Hybrid Deoxyribonucleic Acid," *Nature* 199 (1963): 1034–1040.
146 Holliday's suggestion—R. A. Holliday, "A Mechanism for Gene Conversion in Fungi," *Genetical Research* 5 (1964): 282–304.
147 Lindahl's statement—T. Lindahl, "DNA Repair Enzymes," *Annual Review of Biochemistry* 51 (1982): 61–87.
147 Meselson—Conversation with Matthew Meselson, September 15, 1995.
152 Meselson's paper presented at the 16th International Congress of Zoology, 1963—M. Meselson, "The Duplication and Recombination of Genes," in *Ideas in Modern Biology*, ed. J. A. Moore (Garden City, New York: Natural History Press, 1965), pp. 3–16.
153 Lindahl—Correspondence with Tomas Lindahl, October 13, 1995.
154 Wildenberg and Meselson's results—J. Wildenberg and M. Meselson, "Mismatch Repair in Heteroduplex DNA," *Proc. N. A. S.* 72 (1975): 2202–2206.
154 "prophetic insight about the correction of mismatches"—R. Wagner and M. Meselson, "Repair Tracts in Mismatched DNA Heteroduplexes," *Proc. N. A. S.* 73 (1976): 4135–4139.
155 Meselson's review—M. Meselson, "Methyl-Directed Repair of DNA Mismatches," in *The Recombination of Genetic Material*, ed. K. B. Low (New York: Academic Press, 1988), pp. 91–113.
158 results of experiments with *dam* mutants—M. Radman, R. E. Wagner, Jr., B. W. Glickman, and M. Meselson, "DNA Methylation, Mismatch Correction and Genetic Stability," in *Progress in Environmental Mutagenesis*, ed. M. Alečević (Amsterdam: Elsevier/North Holland Biomedical Press, 1980) pp. 121–130.

CHAPTER 6

161 Witkin's review—E. M. Witkin, "Ultraviolet Mutagenesis and the SOS Response in *Escherichia coli*: A Personal Perspective," *Environmental and Molecular Mutagenesis* 14 (1989): Suppl. 16, 30–34.
161 "It became reasonable to think of UV mutagenesis as error-prone replication..." —Quote from Witkin's review cited above.
163 first observation of prophage induction—Horace Freeland Judson, *The Eighth Day of Creation: Makers of the Revolution in Biology* (New York: Simon and Schuster, 1979), p. 373.
163 Delbrück—Ernst Peter Fischer and Carol Lipson, *Thinking About Science: Max Delbrück and the Origins of Molecular Biology* (New York: W.W. Norton, 1988), pp. 189–190.
163 Festschrift for Delbrück—Judson, *The Eighth Day of Creation*, pp. 374–375.
164 Jacob's recollection—Judson, *The Eighth Day of Creation*, p. 385.
165 Witkin's suggestion—E. M. Witkin, "The Radiation Sensitivity of *Escherichia coli* B: A Hypothesis Relating Filament Formation and Prophage Induction," *Proceedings of the National Academy of Sciences* 57 (1967): 1275–1279.

165 Witkin—Conversations with Evelyn Witkin, May 22 and July 25, 1995.
167 Clark—Conversation with John Clark, August 9, 1995.
167 Witkin's introductory lecture at the conference in Saclay, France in 1990—E. M. Witkin, "RecA Protein in the SOS Response: Milestones and Mysteries," *Biochimie (Paris)* 73 (1991): 133–141.
168 Brooks and Clark—K. Brooks and A. J. Clark, "Behavior of λ Bacteriophage in a Recombination Deficient Strain of *Escherichia coli*," *Journal of Virology* 1 (1967): 283–293.
168 letter by Fuerst and Siminovitch—C. R. Fuerst and L. Siminovitch, "Characterization of an Unusual Defective Lysogenic Strain of *Escherichia coli* K12(λ)," *Virology* 27 (1965): 449–451.
168 paper by Brooks and Clark—Cited previously, noted above.
169 Devoret—Conversation with Raymond Devoret, September 24, 1995.
170 Witkin's hypothesis for the UV nonmutability of *exr* mutants—E. M. Witkin, "Mutation-proof and Mutation-prone Modes of Survival in Derivatives of *Escherichia coli* B Differing in Sensitivity to Ultraviolet Light," *Proceedings of the Brookhaven Symposia in Biology* 20 (1967): 17–55.
170 *recA* mutant is not mutable by exposure to UV light—E. M. Witkin, "The Mutability toward Ultraviolet Light of Recombination-deficient Strains of *Escherichia coli*," *Mutation Research* 8 (1969): 9–14.
171 Witkin's historical reflections on the emergence of the SOS phenomenon—Cited previously, note to p. 161.
172 Radman—Conversation with Miroslav Radman, October 5, 1995.
173 Weigle's finding—J. J. Weigle, "Induction of Mutations in a Bacterial Virus," *Proc. N. A. S.* 39 (1953): 628–636.
174 UV irradiation of both phage and host is necessary for Weigle reactivation—Cited previously, note to p. 173.
174 paper by the Brussels group—M. Defais, P. Fauquet, M. Radman, and M. Errera, "Ultraviolet Reactivation and Ultraviolet Mutagenesis of λ in Different Genetic Systems," *Virology* 43 (1971): 495–503.
175 paper providing evidence for an inducible, error-prone DNA repair mechanism—J. George, R. Devoret, and M. Radman, "Indirect Ultraviolet-reactivation of Phage Lambda," *Proc. N. A. S.* 71 (1974): 144–147.
179 papers on the *tif* (T-44) mutant—M. Castellazzi, J. George, and G. Buttin, "Prophage Induction and Cell Division in *E. coli*. I. Further Characterization of the Thermosensitive Mutation *tif-1* Whose Expression Mimics the Effects of UV Irradiation," *Molecular & General Genetics* 119 (1972): 139–152; M. Castellazzi, J. George, and G. Buttin, "Prophage Induction and Cell Division in *E. coli*. II. Linked (*recA zab*) and Unlinked (*lexA*) Suppressors of *tif-1* Mediated Induction and Filamentation," *Mol. Gen. Genet.* 119 (1972): 153–174.
180 Witkin's paper at the Bethesda meeting in October 1972—E. M. Witkin and D. L. George, "Ultraviolet Mutagenesis in *polA* and *uvrA polA* Derivatives of *Escherichia coli* B/r: Evidence for an Inducible Error-prone Repair System," *Genetics* Suppl. 73 (1973): 91–108.
180 Witkin's paper at the XIII International Congress of Genetics, Berkeley, California, 1973—E. M. Witkin, "Elevated Mutability of *polA* Derivatives of *Escherichia coli* B/r at Sublethal Doses of Ultraviolet Light: Evidence for an Inducible Error-prone Repair System ('SOS Repair') and Its Anomalous Expression in These Strains," *Genetics* 79 (1975): 199–213.
181 paper by Defais et al.—Cited previously, note to p. 174.
181 Radman's paper at the Rochester meeting—M. Radman, "Phenomenology of an Inducible Mutagenic DNA Repair Pathway in *Escherichia coli*: SOS Repair

Hypothesis," in *Molecular and Environmental Aspects of Mutagenesis,* ed. L. Prakash, R. Sherman, M. Miller, C. Lawrence, H. W. Tabor (Springfield, Illinois: Charles C. Thomas, 1974), pp. 128–142.

182 Witkin's 1982 paper—E. M. Witkin, "From Gainesville to Toulouse: The Evolution of a Model," *Biochimie (Paris)* 64 (1982): 549–555.

182 Witkin's comments on the Gainesville meeting—Excerpt from Witkin's paper cited above.

CHAPTER 7

185 collective cellular DNA repair machinery is "Molecule of the Year" 1994—D. E. Koshland, "Molecule of the Year: The DNA Repair Enzyme," *Science* 266 (1994): 1925; E. Culotta and D. E. Koshland, "Molecule of the Year: DNA Repair Works Its Way to the Top," *Science* 266 (1994): 1926–1929.

185 Judson's book—Horace Freeland Judson, *The Eighth Day of Creation: Makers of the Revolution in Biology* (New York: Simon and Schuster, 1979). As noted on p. 191, the expanded edition is now available (Cold Spring Harbor, New York: Cold Spring Harbor Laboratory Press, 1996)

Index

Acridine orange, 73
Acriflavine, inhibition of excision repair, 71
ada, 134–135
Ada, 134, 137
    as suicide protein, 135–136
"Adaptive response" to alkylating damage, work of Cairns and Samson, 134
Agarose gel electrophoresis, 127
Aldous, Elaine, 17, 165
    importance to history of radiation biology, 74–75
alkA, 134
AP endonucleases, 122, 125, 127
Astrachan, Larry, work on genetics of *rII* region of phage T4, 78–79
Ataxia telangiectasia, 99, 188
Atomic Energy Commission, 6–7, 52, 98
ATP-dependent restriction endonuclease in *E. coli* K12, 153
Avery, Colin, 6
Avery, Oswald, 10, 49, 140
    classic 1944 paper with McCarty showing genetic material as DNA, 4–5, 9
    work on pneumococcal transformation phenomenon at Rockefeller, 5
Auxotroph/prototroph method for mutant selection, 69

Bacteriophage, 163
    λ, lysogenic induction, 124
    work of Luria and Delbrück, 22
    UV radiation, effect on, 22
Base excision repair, 116, 122, 126–128
    discovery by Lindahl, 119
    pathway, 118
Beadle, George, at Caltech, 55
Belling, John, work on copy choice replication, 151
Benzer, Seymour, 114
    maps structure of *rII* of phage T2 by mutational analysis, 54
Bertani, Giuseppe, multiplicity reactivation of X-irradiated phage, 23
Beukers, Berends, and IJlstra, UV absorption of frozen vs. unfrozen thymine samples, 58, 80
Blanco, Manuel, 174–175
Bloom, William, 72
Bloom's syndrome, 188
Blunt and Downes report that sunlight kills bacteria, 2
Bohr, Niels, 12–14, 172
Bollum, Fred, 7, 81–82
Borek-Ryan effect, 175
Bovie, William T., localization of UV radiation action by selective absorption, 17
Boyce, Richard, 66, 84–85, 92–93, 167
    generality of excision repair, documentation, 101
    graduate work with Setlow at Yale, 91
Brandeis University, 116
Brenner, Sydney, 54, 150
Bresch, Carsten, 56
Bridges, Bryn, 70

Bromouracil, 54, 89, 118
Bronk, Detlev, 139–140
Brookhaven National Laboratory, Lacks at, 141
Brooks, Katherine, 167–168
Brown, Dan, 119
Burton, Alan C., mentor of Haynes, 72

Caffeine, inhibition of excision repair, 71
Cairns, John, 47, 145, 174
    autoradiography technique, 129
    contributions to DNA repair field, 129
    director of
        Cold Spring Harbor Laboratory (CSHL), 130
        Imperial Cancer Research Fund (ICRF), 131
    views on target theory of excision repair, 75
Caltech, 50, 89, 150, 173
    Delbrück's years at, 15–16, 21, 50, 55
Campbell, Alan, 166, 168
Carrier, Bill, 66, 94, 118
    work with Setlow on excision repair, 81–84
Case Western Reserve University, 78–79, 115, 123–124, 140–141, 166
Chargaff, Erwin, discovery of base ratios in DNA, 123
Chase, H.B., postirradiation recovery phenomenon, 73
Chase, Martha, 51
Clark, Alvin (John), 166
Cleaver, James, 98
Cleland, Ralph, 22
Cockayne syndrome, 187–188
Cohen, Seymour, 81, 88, 114
Cold Spring Harbor Laboratory (CSHL), 2–4, 8, 29–30, 68–69, 115, 129–130, 162
    Symposia on Quantitative Biology, 1–3, 6, 8, 78
Colon cancer
    relationship between mismatch repair and, 185
Columbia University, 22, 67, 140
Complementarity, as suggested in Bohr's lecture "Light and Life," 12
Copy choice replication, 151
Crick, Francis, 53, 123, 150
    Watson/Crick model of DNA, 4, 9, 48–49, 63
    What Mad Pursuit: A Personal View of Scientific Discovery, 49
Crossing-over theory, 145–146
Curie Institute, 156
Curtis, John, work with Hollaender on radiation damage recovery in irradiated bacteria, 18–19
Curtis, Roy III, 130

dam, 157
    isolation, work of Marinus and Morris, 155
Dana Farber Cancer Center, 127
Daniels, Farrington, 18
Dark
    reactivation process, 92
    repair. *See* Excision repair
Darlington, Cyril, 14
Dee Margulies, Ann, work with Clark on recombination, 166–167

203

Deinococcus radiodurans, work of J. Setlow, 60
Deisenhofer, Johann
  crystal structure of *E. coli* photoreactivating enzyme, 61
  Nobel Prize, 61
"Delayed toxicity," after treatment with alkylating agents, 126
Delbrück, Max, 3, 9–11, 21, 23, 40, 45, 68, 95, 129, 163
  advice to Meselson about graduate work, 149–150
  biography by Fischer and Lipson, 11
  course in photobiology at Caltech, 89
  death in 1981, 14
  demeanor at scientific seminars, 15
  influence of Bohr's lecture "Light and Life" and complementarity, 12
  graduate work at University of Göttingen, 11
  letter to Rockefeller Foundation (1936) proposing research in U.S., 14
  Nobel Prize, 22, 42
  oversees new institute in Cologne, 54–55
  phage group. *See also* Phage, group
    founding with Luria, 19, 22
    "phage treaty," 23–24
  postdoc with Bohr, 12–14
  quantum model of gene mutation, 12
DeLucia, Paula, 131
Demerec, Milislav, 3, 14, 68–70
  director of CSHL 1941–1960, 29
  involvement in mass-production of penicillin, 29
  work with Kelner, 30–33
Devoret, Raymond, 156, 169, 171–173, 183
  Charles Leopold Meyer Prize, 176
  comments on Radman's hypothesis explaining W reactivation and mutagenesis, 176
Diplococcus pneumoniae, 143
  work of Avery, 47–48
Djordjević, B., 97–98
DNA
  alkylation damage, 126, 133
    preventase vs. repairase models, 135
  complementarity of DNA strands, 64
  cross-links, 117
  damage caused by UV radiation and skin cancer, 27
  -dependent protein kinase, 189
  dimerization/monomerization experiments in pure DNA, 59
  dimers as lesions, work of the Setlows, 59–60
  as genetic vs. purely structural function, 9
  glycosylases, 117, 119, 121–122, 125, 127, 137
  hemimethylated, 146
  heteroduplex molecules, work of Meselson and Wildenberg, 156
  methylation, 155
    role as basis for strand discrimination during mismatch repair, 157–158
  polymerase, 66, 83, 121
    I, 130, 131
    II and III, 131
    assay, 131
  recombination, alkylation damage in DNA, 124
  repair, 17–19, 134, 142. *See also* Enzymatic photoreactivation; Excision repair
    comments by Luria on multiplicity reactivation, 22–23
    in *E. coli*, 134–138
    factors influencing delayed emergence of field, 11
    field born by efforts of Kelner, Dulbecco, Rupert,
and Goodgal, 53
    first meeting dedicated to field, 96
    Hollaender as founding father of field, 6
    and human disease, 99
    and mutagenesis, 2, 70–71
    and transcription, 188
    vs. excision repair, 63
  replication, 139
    associated with phage recombination, work of Meselson and Weigle, 152
  sequencing gels, 127
  spontaneous hydrolysis, 120
  stability
    comments by Luria, 22
    comments by Mazia in 1952, 17
  synthesis
    and excision repair, 63
    inhibition, 47
  transformation, 140–143
    of *S. pneumoniae*, 141
  UV-irradiated, formation of cytosine and thymine dimers, 58
Dobzhansky, Theodosius, mentor to Witkin, 78
Doudney, Charles, 70
Downes and Blunt, report (1877) that sunlight kills bacteria, 2
Drakulić, Maria, 172
Drosophila studies
  foundation for early field of genetics, 3–4
  X rays enhance mutation frequency, 1–2, 11–12, 21
Duggar, Benjamin, 18
Duke University, 77, 144
Dulbecco, Renato, 51, 129
  discovery of photoreactivation in phage, 24, 28, 37
  effect of temperature, 39
  Nobel Prize in 1975, 42
  UV-inactivated phage reactivated by visible light, 36
  work with Luria
    on cellular responses to DNA damage, 20
    joins Luria's laboratory, 42
Duncan, James, 122

Echols, Harrison, 180
The Eighth Day of Creation: Makers of the Revolution in Biology (Judson), 1, 5, 78, 123, 150
Ellis, Emory, 14
Emmons, C.W., 6
Endonuclease II/VI, 122, 125–126
Enzymatic photoreactivation
  first form of DNA repair discovered, 27–29
Enzyme-catalyzed DNA repair reaction, first indication, 50
Ephrussi, Boris, 78
Ephrussi-Taylor, Harriet, 78, 140–141, 143, 146, 153
Errera, Maurice, 172, 174
Escherichia coli, 131
  B strain
    B/r, 81–82. *See also* Witkin, Evelyn
    B_{s-1}, 54–55, 65, 81–82. *See also* Hill, Ruth
    exposure to UV radiation, 54
    PhR-effect, 57
  circular chromosomes, caught in the act of replication by autoradiography, 130
  DNA repair, 134–138, 142. *See also* DNA repair
  excision repair, 94
    enzymes, 127
    in UV-irradiated *E. coli*, 83

gene regulation, 149
induction of expression on β-galactosidase, 149
K12 strain, 153, 163
  uvrA, *uvrB*, and *uvrC* isolation, 92
  mechanism of UV-radiation-induced mutagenesis, 156
  mismatch excision repair and, 154
  photoreactivating enzyme
    crystal structure, 61
    properties vs. *Saccharomyces*, 57
  phr+ encodes photoreactivating enzyme in, 53
  SOS repair. *See* SOS pathway
  study of synchronous growth, 87
  UV-radiation-induced filamentation, 162–163, 179
  WP2s strain, defective in excision repair, 161
Excision repair (dark [photoreactivating light-independent] repair), 186. *See also* Base excision repair
  defect in humans, xeroderma pigmentosum, 98–99
  DNA synthesis and, 63
  as evolutionarily conserved mechanism, 97
  gene *ERCC1*, 188
  generality of, documentation by Boyce, 101
  in mammalian cells, 97
  milestone work of Haynes, 100
  multi-target requirement, 74
  process, 63
  of pyrimidine dimers in *E. coli* exposed to UV radiation, 142
  thymine dimers. *See* Thymine dimers
  of uracil in DNA, 122
  vs. DNA repair, 63
  work of Haynes and Patrick, 73–76
  work of Hill, 54–55, 64
  work of Kelner, 45
  work of Witkin, 67
Exonuclease III, 127
exr mutation, sensitivity to UV radiation and X rays, 170, 174

Fano, Ugo, 20
Fermi Award from the U.S. Dept. of Energy
  to Fano, 20
  to Hollaender in 1983, 6–7
Fermi, Enrico, 21
  work with Luria, 20
Filamentation, UV-radiation-induced, in *E. coli*, 162–163, 179
Fischer, Ernst Peter, biography about Delbrück, 11, 24
Fisher, R.A., 14
Fleming, Alexander, discovery of penicillin, 28
Fluke, Donald, 80
Fox, Maurice, study of DNA transformation, 143–144
Fresco, Jacques, 119
Friedberg, Errol C.
  discovery of phage T4 v^+ gene product, 116
  drafted into U.S. Army, 85
  first meeting with R. Setlow in 1966, 78
  friendship with Radman, 171
  move to Stanford in 1971, 113
  postdoc in Goldthwait's laboratory, 124
  work on enzyme activity from phage T4-infected cells, 113
Freifelder, David, 72
Freese, Ernest, 54

Ganesan, Ann, 117–118

assay for detection of dimers, 87
Garen, Alan, 64–65
Gates, Frederick, 8
Gatlinburg Symposia, 7, 155
Gene
  function, spontaneous alterations/mutations, 1
  nucleic acid composition as proposed by Hollaender and Emmons, 8
  regulation. *See* Prophage induction
George, Donna, work with Witkin, 180
George, Jacqueline, 176, 180
Gilbert, Walter, isolation of Lac and λ repressors, 164
Glickman, Barry, 158
Goldthwait, David, 78–79, 85, 115, 123–124, 129
  sabbatical at Pasteur Institute, 166
  work on SOS phenomenon, 162, 165, 168
Goodgal, Solomon, 78
  demonstration that photoreactivation is DNA repair process, 49–50
  interest in DNA transformation, 51
Gothlin, John, 148
Grafstrom, Robert, 117
Greenberg, Bill, isolation and characterization of deoxyribonucleases, 143
Grossman, Lawrence, 127
  and Delbrück, 56, 117
  studies on photochemistry of nucleic acids, 116
Gudas, Lorraine, model for SOS response, 183
Guiduschek, Peter, 72
  demonstration of recombination between *B. subtilis* and *M. lysodeikticus*, 115–116
Guild, Walter, 75, 144
  Radiation Research paper (1963), 77

Haas, Felix, 70
Hadi, Sheik, 125
Haemophilus influenzae
  photoreactivation, 52–53
  transformation with DNA, 52
  work of J. Setlow, 60
Haldane, J.B.S., 14
Hammond, George, 89
Hanawalt, Philip C., 79, 81, 85–86, 117, 187
  letter to Setlow describing discovery of excision repair, 90
  paper with Haynes (1964), 64
  Stanford University faculty in 1961, 89, 100
  study of synchronous growth in *E. coli* with Morowitz, 87
  work
    on DNA repair, 63, 66, 74
    in Setlow's laboratory, 88
Harm, Walter, 23, 65, 75, 82
  and Helga at Southwest Institute, 50
  isolation of UV-sensitive mutant of *E. coli*, 55, 57
  PhR
    -effect with *E. coli* B, 57
    enzyme, 55
  u-gene reactivation, 24, 65–66
Harris, Reginald, director of CSHL and initiator of Symposia on Quantitative Biology, 3
Harvard University, 133, 141
Harvey Society, lecture given by Delbrück in 1946, 14
Haselkorn, Robert, 72
Haseltine, William, 117
Hausser, Rolf, 41
Haynes, Robert H., 10, 63

discovery of liquid holding recovery in yeast, 73–76, 100
paper with Hanawalt (1964), 64
speculation of DNA repair mode in 1960s, 16
training at McGill University, 72
work with Zirkle, 72
hcr mutation, UV-sensitive phenotype, 170. *See also uvrB*
Herriot, Roger, 50, 52, 56, 116
 letter to Hershey describing phage infection of bacteria, 50–51
Hershey, Alfred, 16, 50–51, 129, 150
 Nobel Prize, 22, 42
Hershey-Chase Waring blender experiments, 51, 150
Hertel, Ernst, 41
Hertman, Israel, work on *recA*, 165
hex mutants, mismatch repair and, 143, 154
Hill, Ruth, 68, 80, 92, 161, 170, 172
 discovery and elucidation of excision repair of DNA, 64–65
 host cell reactivation, 64
 isolation of UV-radiation-sensitive mutant of *E. coli* (B_{s-1}), 54–55, 65, 161
Hillebrandt, Brigitte, work with Harm in Cologne, 57
Hollaender, Alexander, 3, 7, 38, 64, 67, 70, 74, 80
 background, 6
 as director of Biology Division at Oak Ridge National Laboratory, 7
 Enrico Fermi Award, 6–7
 founding father of DNA repair field, 6
 "History of Radiation Biology from a Personal Point of View," 6
 National Research Council Project in Wisconsin, 18
 presentation at 1941 CSHL Symposium, 8
Holliday, Robin, 145–146, 153
 speculation about mechanism of gene conversion, 145
"Holliday junction," 145
Hotchkiss, Rollin D., 139–140, 143
Howard-Flanders, Paul, 66, 84–85, 90, 153, 167, 169, 171
 discovery of postreplication/recombinational repair, 91
 excision repair in *E. coli*, vs. work of Setlow, 94
 K12 *uvrA*, *uvrB* and *uvrC* isolation, 92
Human Genome Project, 42
Hutchinson, Frank, 75
Hutchinson, Wesley G., mentor to Kelner, 34

Inch, W.R., reactivation mechanisms, work with Haynes, 77
Indiana University, 1940s leading center of genetics, 22
Institut für Genetik in Cologne, 65
 established by Delbrück, 55
Ionizing radiation. *See also* X rays
 early theories about mutations and chromosomal aberrations, 4

Jacob, François, 124, 149, 164, 166
Jagger, John, 8, 50, 80
Johannsen, Wilhelm, description of mutation, 11
Johns, Harold, 89
Johns Hopkins University, 50–51, 53, 56, 116, 126
Judson, Horace, 9, 54, 140, 152, 163
 The Eighth Day of Creation: Makers of the Revolution in Biology, 1, 5, 78, 123, 150

Kaplan, Henry S., 89
Karolinska Institute, 118
Kaufman, Berwin, 4
Kelner, Albert, 40–41, 43, 47, 56, 70, 74
 background, 28
 discovery of photoreactivation, 28–30
  letter to Luria, recovery after UV irradiation and exposure to visible light, 34–35
  light exposure vs. temperature, 34
  temperature-dependent recovery hypothesis, 32–34
  UV-irradiated cell survival and logarithmic relationship to radiation dose, 45
 move to CSHL to work with Demerec, 30
 search for new antibiotics while at U. Penn., 29
Kikkawa, Hideo, 115
Kilby, B.J., 16
Kimball, Richard
 work distinguishing true gene mutations, 4
Kimball, William, 7
King, Jack, 86, 113, 115–116
Kirby, Edward, work with a filamenting strain of *E. coli*, 162, 166, 168
Korn, David, 85
Kornberg, Arthur, 49, 51–52, 79
 discovery of DNA polymerase, 66
 exonuclease III, 127
Korogodin, V.I., liquid holding recovery, 73–74
Krieg, David, 90
Ku autoantigens, 189
Kyushu University, 114

Lacks, Sanford, 139–140, 144, 153, 155
 work with Monod, 141
Latarjet, Raymond, postirradiation recovery of bacteria and cold, 19
Laval, Jacques, 125
Lawly, Philip, 126
Lawrence, Ernest O., request to build cyclotron at Yale, 79
Lederberg, Joshua, 9, 99
Lehman, I. Robert, 52
 friendship with Goodgal, 51
Levinthal, Cyril, 53
Levi-Montalcini, Rita, 42
lex, 170
 designation by Howard-Flanders, 169
lexA, 179, 181
 required for Weigle reactivation, 174
LexA, role in repression of SOS genes, 161, 168–169, 171, 183
Lieb, Margaret, 70
Lindahl, Tomas, 118, 125, 127, 134
 discovery of DNA glycosylation and base excision repair, 119
 mechanism of adaptation to alkylation mutagenesis, 136–138
 review of DNA repair enzymes, 147
 uracil-DNA glycosylase, 121
Lipson, Carol, biography about Delbrück, 11, 24
Liquid holding recovery in yeast, discovery by Haynes and Uretz, 73, 100
Lotte, Charlotte Auerbach, work with mustard gas as mutagen, 16
Loveless, Tony, 135
Luria, Salvador, 3, 10, 21, 34–36, 54, 65, 129, 165
 comments on career in medicine, 20

earliest documented use of term "repair," 48
founding of phage group with Delbrück, 19, 22. *See also* Phage, group
influenced
 by Delbrück's notion of gene as molecule, 21
 by Fano, 20
multiplicity reactivation, 22, 64
 of UV-irradiated phage, 40
Nobel Prize, 22, 42
A Slot Machine, A Broken Test Tube, 19–20, 22
Lwoff, André, work on prophage induction, 163
Lysenko, Trofim, 67
Lysogeny, 163

Maaløe, Ole, 88
MacLeod, Colin, paper showing that genetic material was DNA (1944), 4
Maize, studies and foundation for early field of genetics, 3–4
Maltose-utilization marker, 140–141
Manhattan Project, 46, 68, 72
Marine Biological Laboratory at Woods Hole, 150
Marinus, Martin, isolation of *dam*, 155
Markert, Clem, 50
Marmur, Julius, 116
Massachusetts Institute of Technology, 143, 168
Mazia, Daniel, 3
 comments on stability of DNA, 17
 "the guru of mitosis," 16–17
McCarty, Maclyn, 140
 classic 1944 paper with Avery showing genetic material as DNA, 4, 9
 The Transforming Principle: Discovering that Genes are made of DNA, 9–10
McClintock, Barbara, 3, 4
McGill University, 72
Meselson, Matthew, 67, 129–130, 147, 166, 173
 decision to
  become a scientist, 147
  train with Pauling, 148
 graduate school encounter with Delbrück, 149
 ideas for investigating mechanism of DNA replication, 149–151
 meets
  Stahl at Woods Hole in summer 1954, 150
  Watson at Caltech, 150
 move to Harvard, 152
 scorn of phage group, 21
 -Stahl DNA replication experiment, 17, 21, 129
 sucrose gradient sedimentation experiments
  addition of ATP, 153, 155
  preliminary experiments, 149
 work
  on crystal structure of tellurium, 148
  with phage vs. bacteria, 21
O^6-Methylguanine
 -DNA methyltransferase, 138
 and O^4-methylguanine in DNA, role in mutagenesis, 135–136
 repair in vitro, 137
Micrococcus
 Deinococcus radiodurans, work of J. Setlow, 60
 luteus, UV-radiation-resistant bacterium, 24, 115
Mirsky, Alfred, 3, 10
Mismatch repair, 146
 description, 139
 genes, 188
 mechanism of strand discrimination, 146
 as mutational avoidance mechanism, 139–140
 in pneumococcus, 141
 relationship between mismatch repair and colon cancer, 185
 role of DNA methylation, 152, 158
Mitomycin C, lethal effects and excision repair, 101
MNNG (*N*-methyl-*N'*-nitro-*N*-nitroso-guanidine), as mutagen, 133–135
Modrich, Paul, work on mechanism of mismatch repair, 158
Molecular biology
 emergence, 24–25
 term coined by Weaver, 6
Monochromators, 91
 use by Setlow to determine action and absorption spectra for DNA, 80
Monod, Jacques, 54, 141, 164, 176
 induction of expression on β-galactosidase in *E. coli*, 149
Morgan, Thomas Hunt, study of spontaneous mutations in *Drosophila*, 1
Morowitz, Harold, 75, 87
Mosbaugh, Dale, purification and characterization of phage PBS2 inhibitor, 118
Muller, Hermann J., 3, 9, 22
 work with Auerbach Lotte on radiation-induced mutations in *Drosophila*, 16
 X rays enhance mutation frequency in *Drosophila*, 1–2, 11–12, 21
Multiplicity reactivation, discovery by Luria, 22–23
Mustard gas, as alkylating agent, and ionizing radiation, 16–17
mut, defective in mismatch repair, 158
Mutations/mutagenesis
 "delayed effect," 69, 71
 description in 1909 by Johannsen, 11
 induction mechanisms in early 1940s, 4
 O^6-Methylguanine and O^4-methylguanine in DNA, mutagenesis role, 135
 molecular mechanism for spontaneous mutations, 48–49
 mutation frequency decline, work of Witkin, 69–71
 natural vs. experimental, early work, 1

Nadson, G.A., "radiation sickness," 17, 73
Nathans, Dan, work with restriction enzymes, 151–152
National Laboratories for radiation research. *See also* Atomic Energy Commission; Oak Ridge National Laboratory
 established by President Truman, 6
Nethybridge meeting, importance to mismatch repair, 154–155, 157
Nitrogen mustard, as alkylating agent, exposure and liquid holding recovery, 100
Nitrous acid, 54, 65
 -induced mutagenesis, 57
Nobel Prize, 22, 42, 61, 120
Northrop, John Dexter, 6
nov marker, 144
Novick, Aaron, 46, 72, 143

Oak Ridge National Laboratory, 7, 17, 45, 58–59, 66, 78–80, 98, 131
Oehme, Reinhold, 41
Okubo, Shunzo, 114
Olsson, Monica, 138

Orgel, Leslie, 49
Osaka University, 115

Painter, Robert, 98
  work on excision repair in mammalian cells, 96–97
Paquette, Yves, 126
Paramecium caudatum, study of UV radiation by Bovie, 17
Pasteur Institute, 166, 173
Patrick, Michael, 74
  dye-sensitized photodynamic inactivation system, 72
Patt, Harvey, 98
Pauling, Linus, 7, 148, 149
Pettijohn, David, 66, 85, 89, 100
Phage. *See also* Bacteriophage
  λ
    host modification, 152
    recombination by breakage and rejoining, 151
    repressor, 162
  group, 9
    contributions to molecular biology, 19
    with Delbrück at Caltech, 15
    founding by Luria and Delbrück, 19, 22
  lysogenic induction, 175, 182
  mutagenesis, effects of UV radiation, 47
  P1, 153
  PBS2, 121
    infection of *B. subtilis*, 118
  radiation genetics, 150
  reactivation, 64. *See also* Photoreactivation
  T1, UV irradiation, 92
  T4, 84–86
    denV endonuclease, 113
    pyrimidine-dimer-specific enzyme, 115
    resistance to inactivation by UV radiation, 24
  T4v1, 86
  T7, 126
"Phage treaty," 23–24
Phillips, Ted, 97
Photoreactivation, 44, 76, 90, 128
  decrease in number of mutants, 46
  Delbrück's comments to Luria, 46
  as DNA repair process, 46, 49–50
  first demonstration in vitro, 50
  of *Haemophilus influenzae*, 52–53
  incandescent bulbs vs. fluorescent light, 38
  limitation, 47
  photoreactivating enzyme, 61, 117
  substrate of photoreactivation is DNA, proof using transformation assay, 52
  ultimate demonstration by Goodgal and Rupert that photoreactivation is DNA repair process, 49–50
  work of Dulbecco and Luria, 24
  photoreactivation of UV-irradiated phage T2, 32
  work of Kelner, 33–38
Pneumococcal transformation phenomenon, 140
  integration efficiency, 144
  work of Avery, 5
Pneumococcus type III, 5
polA, 131, 179
Pollard, Ernest, 79, 87–88, 183
Postreplication/recombinational repair, discovery by Howard-Flanders, 91, 167
Preventase, 135
Princeton University, 119
Pronase, 138

Prophage induction, 163–164
Ptashne, Mark, 176
  isolation of Lac and λ repressors, 164
Purdue University, 114
Pyrimidine dimers, 63, 114, 116–117, 161, 177. *See also* Photoreactivation
  absolute specificity of photoreactivation for, 100
  discovery of *M. luteus* pyrimidine primer DNA glycosylation activity, 128
  interaction of UV radiation with DNA and, 27–28
  removal by excision repair and loss of photoreactivation, 45

Radany, Eric, work with phage T4 UV endonuclease, 128
Radiation
  cell mutation frequency with radiation exposure, 4
  ionizing, 16–17. *See also* X rays
    early theories about mutations and chromosomal aberrations, 4
  sickness, work of Nadson, 17
  UV radiation. *See* UV radiation
  Weaver's report on biological effects of radiation, 6
  work at National Laboratories dedicated to radiation research, 7
Radiobiology, progression to molecular biology, 20–21
Radman, Miroslav, 144, 156, 162, 176
  Charles Leopold Meyer Prize, 176
  introduction to phage λ, 172–173
  postdoc in Devoret's laboratory, 171
  purification of putative mismatch repair endonuclease, 157
  role in mismatch repair story, 154–155
  SOS pathway work, 177–181
    confirmation of SOS hypothesis by Witkin, 179–180
    early formulation of basic SOS model, 175
  work on mechanism of Weigle reactivation, 174–175
Rasmussen, Ronald, excision repair in mammalian cells, 96, 98
recA, 171, 176, 181
  dependence on filamentous growth, 179
  induction of phage λ, 165–166, 169
RecA, 162, 167–168, 183
Recombination, 151, 170, 176. *See also* Mismatch repair
  alkylation damage in DNA, 124
  between *B. subtilis* and *M. lysodeikticus*, 115–116
  biological role, and mismatch repair, 154
  as mechanism of DNA repair, 76
  of phage
    λ DNA, 151
    T4 and T7, 167
  recombinational repair, 91
Repairase model of DNA alkylation damage, 135
Restriction enzymes
  type I, 151
  type II, 151–152
Richardson, Charles, discovery of exonuclease III, 127
Riklis, Emmanuel, 94
Ris, Hans, 3
Rita, Geo, 21
Ritter, J.W., discovery of UV radiation in 1801, 2
RNA polymerase II, 185, 187
Roberts, Richard B., 165
  importance to history of radiation biology, 74–75
  work with Aldous on *E. coli* exposure to UV radiation, 17

Robins, Peter, 135–136
Rockefeller Foundation, 13–14
Rockefeller Institute, 18, 139
Rockefeller University, 5–6
  Avery's work on pneumococcal transformation phenomenon, 5
  Gates's work on radiation and bacteria, 8
  Lederberg as President, 9
  Weaver's report on biological effects of radiation, 6
Roller, Ann, 172
Rupert, Claud S. (Stan), 28, 30, 55–56, 59
  assay for DNA as substrate in photoreactivation, 52
  enzymatic photoreactivation as life work, 53
    equilibrium reaction, 57
    loss of transforming activity by heating and dialysis, 53
    photoreactivating enzyme, 61
  intrigue with DNA, 51
Rutchevsky, Nicholas, 148

Saccharomyces cerevisiae, 72
  liquid holding recovery, 73
  photoreactivating enzyme activity, vs. *E. coli*, 57–58, 61
Sakami, Warwick, 124
Samson, Leona, 135
  "topostat" system, 132
Sancar, Aziz, cloning of *E. coli phr*, DNA repair gene, 61
Sancar, Gwendolyn, 61
Sauerbier, Walter, 65
Schendel, Paul, 135
Schrödinger, Erwin, 11, 21, 72
  What is Life? (1944), 13, 79
Sekiguchi, Mutsuo, 86, 114–116, 134
Setlow, Jane, 7, 58–59, 91–92, 94
  contributions to DNA repair field, 60
  Rupert's introduction of Setlow at 1965 Chicago meeting, 59–60
Setlow, Richard B., 7, 58, 67, 71, 77, 79, 90, 94, 153
  determination of
    action spectrum for splitting thymine dimers, 59
    kinetics of dimerization and monomerization, 59
  excision repair work
    with Carrier, 83–85, 92–94
    controversy and Howard-Flanders work, 94–95
    role in discovery, 66
    simultaneous publication with Boyce and Howard-Flanders, 95
  and Hollaender, 7, 58–59
  introduction to thymine dimers, 80
  at Oak Ridge National Laboratory, interest in DNA damage and repair, 80–81
  at Yale as physical spectroscopist, 80
Sherman, F.B., 73
Sinsheimer, Robert, 89
Special Research Aid Fund for Deposed Scholars, 13–14
Sonneborn, Tracy, 22
SOS pathway (SOS regulon/SOS system), 161
  derepression of SOS genes, activation by UV exposure, 162
  evolutionary significance, 183
  LexA, role in repression of SOS genes, 161. *See also* LexA
  origin of term SOS, 177
  RecA, 162
  response to UV radiation, 124
  role of Witkin. *See* Witkin, Evelyn
  transcription of SOS genes, 161–162
  work of Radman, 176–182
Southwest Institute for Advanced Studies, contributions to photobiology, 50, 55
Stadler, Lewis, 3
  work with X-ray-induced mutations in maize, 11
Stahl, Franklin, 147, 149
  comments on why gene/DNA repair was late in coming, 17
  Meselson-Stahl DNA replication experiment, 17, 129
Stanford University, 41, 66, 85, 87, 89
Starlinger, Peter, 56
State University of New York (SUNY), 70
Stern, Curt, 3
Straub, Joseph, 55
Strauss, Bernard S., 66
  discovery of pyrimidine-dimer-specific activity of *M. lysodeikticus*, 115
  memories of Delbrück at Caltech, 15–16
Strauss, Lewis, Chairman of Atomic Energy Commission, 7
Streptococcus pneumonia
  hex mutants, as spontaneous mutators, 154
  type II bacterium, work of Griffith, 5
Streptomyces griseus (UV-irradiated)
  action spectrum for photoreactivating light in, 48
  exposure to light, 33
  UV survival curve, 45
Streptomycin-resistance marker, 140
Sucrose gradient sedimentation, 86, 124–125, 153. *See also* Meselson, Matthew
Suicide proteins, 134–136
sul (*sfi*), 68
Sutherland, Betsy, 83
Swenson, Paul, 17–18, 66
Symonds, Neville, emerging field of genetics in 1940s (summary), 3–4
Szent-Gyorgi, Albert, photochemical reactions in aqueous and frozen samples, 57–58
Szilard, Leo, 40, 46, 68, 143

T-44, 176, 179. *See also tif*
  filamentation and, 166, 168
T70, recombination-defective mutant, 168
Takagi, Yasuyuki, 114
Theriot, Lee, 169
Thomas, René, 174
Thymine dimers, 89
  acid-stability, and DNA photoreactivation, 59
  action spectrum for splitting, 59
  content in whole cells, 82
  controversy and excision repair, 93
  detection using TCA, 94
  role in inhibition of DNA replication, 81
  as substrate for "dark reactivation" process, 92
  UV-irradiated DNA, formation of cytosine and thymine dimers, 58
tif, temperature-sensitive mutant, 179–181. *See also* T-44
Timoféeff-Ressovsky, Nicolai, 13–14
  death in 1981, 14
  director of Genetics Laboratory at Kaiser Wilhelm Institute for Brain Research, 12
  quantum model of gene mutation, 12
Tiraby, Gerard, study of DNA transformation with Fox, 143–144

Todd, Lord Alexander, 120
Tolmach, L.J., 97–98
"Topostat" system developed by Cairns and Samson, 132
Transformation
  pneumococcal transformation phenomenon, work of Avery, 5
  quantitation using streptomycin resistance gene, 52
"Transforming principle," 5, 140
Tufts University, 58

u-gene reactivation, 24, 65–66
Ultraviolet (UV) radiation, 81
  excision repair and, 67
    Haynes' hypothesis for mechanisms for repair of UV radiation and X ray damage, 76
    of pyrimidine dimers, 142
  interaction of UV radiation with DNA and pyrimidine dimers, 27–28
  mechanism of UV-radiation-induced mutagenesis, 179
    in *E. coli*, 156
  as mutagen, 1–2, 17, 23, 30, 54, 161
    action spectrum, absorption by nucleic acids and, 8
    in bacteria, work of Witkin, 68–69. *See also* Witkin, Evelyn
    discovery in 1801 by Ritter, 2
    effect of postirradiation temperature, 32
    effects on microorganisms, work of Hollaender, 8. *See also* Hollaender, Alexander
    method of mutant selection, 54
    monochromatic, use by Hollaender, 8
    post-UV recovery of fungal spore viability, 31
  phage λ initiation, 165
  "poison," 19, 47
  recovery of *E. coli* from UV radiation exposure, 17–18
  as selective agent during biological evolution, 128–129
  -sensitive mutant of *E. coli*, isolation in 1958 by Hill, 54–55
  survival curves, work of Kelner, 31
  UV-irradiated DNA, formation of cytosine and thymine dimers, 58
Union College, 140
Unscheduled DNA synthesis, discovery by Painter, 98
Uracil
  -DNA glycosylase, 121
  incorporation by DNA polymerase during DNA replication, 121
Uretz, Robert, 72, 75
  discovery of liquid holding recovery in yeast, 73
Utter, Merton, 124
uvrA, *uvrB*, and *uvrC*, 101
  isolation by Howard-Flanders and Boyce, 92

v^+ gene of phage T4, 65
Vanderbilt University, 15, 42
Verly, Walter, 125
Volkin, Elliot, 7, 78
von Borstel, R.C. (Jack), 7

Wagner, Robert E., work on biological role of mismatch repair, 154, 157
Waldstein, Evelyn, 94
Walter Reed Army Institute of Research, 83, 85

Wang, Shih Yu, 116
  work on dimerization of uracil in frozen solution, 58
Warner, Huber, 121
Washington University, 6, 51, 66
Watson, James D., 21–23, 32, 43, 88, 130, 149
  arrives in England in 1951, 4
  /Crick model of DNA, 4, 9, 48–49, 63
  director of CSHL, 130
  influence of Shrödinger's book *What is Life?*, 13
  studies on multiplicity reactivation with Luria, 23
Weaver, Warren, 13
  coins term "molecular biology," 6
Weigle, Jean, 151–152, 173
  multiplicity reactivation of X-irradiated phage, 23
  "UV restoration," 64
"Weigle phenomenon" (W reactivation), 171, 173, 177. *See also* Radman, Miroslav
Weiss, Bernard, 126–127
Welch, G. Rickey, 13
Werbin, Harold, purification of photoreactivating enzyme from yeast, 61
Whitaker, Douglas, 41
Whitehouse, H.L.K., 145
Wildenberg, Judy, work with Meselson, 154, 156–157
Williams, Robley C., 79, 87
Willstatter, Richard, study of enzymatic activity in protein-free solutions, 5–6
Witkin, Evelyn, 12–13, 25, 67, 69, 81, 167, 169, 171, 178
  confirmation of Radman's SOS hypothesis, 179–180
  experiments with *tif* and *recA*, 182
  hypothesis for UV nonmutability of *exr* mutant, 170
  importance of term "repair," and SOS phenomenon, 182
  work
    at CSHL with Demerec, 69
    on filamentation of UV-irradiated *E. coli*, 161–162, 165, 169
    at SUNY Downstate, 100
    as young investigator at CSHL in 1940s, 4
  notion of genetic rectification of genetic damage, 4
Wolff, Sheldon, 7
Wood, Harland, 124
Wovcha, Merle, 121
Wulff, Daniel, acid-stability of thymine dimers and DNA photoreactivation, 59

Xeroderma pigmentosum, example of excision repair defect in humans, 98–99, 187
  knock-out mouse strains, 188
X rays
  chromosomal breakage induced by, 4
  inactivation of *Neurospora conidia*, 50
  as mutagens, 2, 18, 23
    Muller's work in *Drosophila*, 1, 11–12
    Stadler's work in maize, 11
  vs. UV radiation, 2, 31

Yale University, 7, 58, 66, 84, 87, 91–92
  cyclotron, 79
York University, 10, 64

Zachau, Heinz, 56
Zickler, H., first demonstration of gene conversion in 1934, 145
Zimmer, K.G., quantum model of gene mutation, 12
Zinder, Norton, 64–65